CALIFORNIA CONDORS

IN THE PACIFIC NORTHWEST

The John and Shirley Byrne Fund for Books on Nature and the Environment provides generous support that helps make publication of this and other Oregon State University Press books possible. The Press is grateful for this support.

Previously published with the support of this fund:

One City's Wilderness: Portland's Forest Park
by Marcy Cottrell Houle

Among Penguins: A Bird Man in Antarctica
by Noah Strycker

Dragonflies and Damselflies of Oregon: A Field Guide
by Cary Kerst and Steve Gordon

Ellie's Log: Exploring the Forest Where the Great Tree Fell
by Judith L. Li
Illustrations by M. L. Herring

Land Snails and Slugs of the Pacific Northwest
by Thomas E. Burke
Photographs by William P. Leonard

California Condors
in the
Pacific Northwest

JESSE D'ELIA AND SUSAN M. HAIG

Illustrations by Ram Papish

Foreword by Noel Snyder

OREGON STATE UNIVERSITY PRESS • CORVALLIS

For Mason and Quinn

May you one day have the pleasure of gazing upward at a sky alive
with the splendid evolutions of the mighty California Condor

The paper in this book meets the guidelines for permanence and durability of
the Committee on Production Guidelines for Book Longevity of the Council on
Library Resources and the minimum requirements of the American National
Standard for Permanence of Paper for Printed Library Materials Z39.48-1984.

Library of Congress Cataloging-in-Publication Data

D'Elia, Jesse.
California condors in the Pacific Northwest / Jesse D'Elia, Susan M. Haig.
   pages cm
  Summary: "The authors study the evolution and life history of the California
Condor, its historical distribution, the reasons for its decline, and their hopes
for its reintroduction in the Pacific Northwest"-- Provided by publisher.
  Includes bibliographical references and index.
  ISBN 978-0-87071-700-0 (pbk.) – ISBN 978-0-87071-701-7 (e-book)
  1. California condor–Northwest, Pacific. 2. California condor–Conservation–
Northwest, Pacific. I. Haig, Susan M. II. Title.
QL696.C53D45 2013
598.9'2–dc23

2012044612

Oregon State University Press
121 The Valley Library
Corvallis OR 97331-4501
541-737-3166 • fax 541-737-3170
http://osupress.oregonstate.edu

Contents

Figures and Tables

Figures

Tables

Foreword

Presently the largest and most astonishing bird in the skies of North America, the California Condor was one of our most highly endangered species by the 1980s, when it persisted only in a region just north of Los Angeles. By the late 1980s it endured only in captivity, but it has since been returned to the wild in selected regions. Fossil evidence from Pleistocene times shows that it inhabited not only California but a continent-wide range stretching from northern Mexico to Florida, New York, and the Pacific Northwest.

The condor was likely a breeding bird in most regions where its fossils have been found, but so far, breeding has been confirmed only in California, Baja California, and Arizona. In Arizona, paleontological research has revealed that the species once nested in caves perforating the many formations of the Grand Canyon and, following releases begun in 1996, it has again returned to nest in these sites. Whether the species ever nested in Oregon and Washington, however, has been a subject of controversy. It was frequently reported seen in this region in the nineteenth century, starting with the epic journey of Lewis and Clark in 1805, but no one has ever documented a contemporary or historic condor nest north of California. This book discusses suggestive evidence that condors were indeed breeders in the Northwest and presents a careful analysis of causes of disappearance of the species from this region by the early twentieth century—efforts that serve as a prelude to a potential program to revive a wild population in the region.

Should a consensus develop that the condor was indeed once a full-time resident and breeder in the Northwest, and should agreement be achieved that the past and present causes of the species' decline in this region have been reliably identified and countered, it may well prove feasible to re-establish this species as a wild creature in the region. This book goes a long way toward justifying such an effort, although it also thoroughly discusses the information gaps and resistance factors still remaining that could prevent success in such a project.

The last wild condor in the remnant historic population in California was trapped into captivity in 1987, joining twenty-six other condors taken from the wild as eggs or otherwise trapped from the wild. Captives have bred readily, and the total captive population, now maintained in part by the Oregon Zoo, has increased rapidly. Numbers of birds have been sufficient to allow the initiation of release programs to the wild in several locations in California, Baja California, and northwestern Arizona. These release populations, which now collectively include several hundred individuals, have been maintained in part on subsidies of carrion food and have all initiated breeding activities. However, none of these populations has yet attained demographic viability because of a variety of problems, the most important of which has been poisoning stemming from the birds feeding on carcasses of hunter-shot game containing lead ammunition fragments.

California passed legislation outlawing the use of lead ammunition in the condor's range in 2007, but the poisonings in this state continue because of difficulties in enforcing the legislation and the wide availability of lead ammunition across the country. It seems likely that an effective solution to the lead-poisoning problem may necessitate national legislation that truly removes the sources of lead ammunition and substitutes other equally effective ammunition that is nontoxic. As lead ammunition also contaminates humans to some extent, especially through ingestion of hunter-shot game, such legislation would also be a significant benefit for human health, to say nothing of the benefits to wildlife species other than condors that also suffer from lead poisoning. The insidious sublethal effects of lead contamination on our own species have already led to a banning of lead in paints, gasoline, and plumbing.

Thus, success in reestablishing condors in the Northwest may well depend on success in national efforts to solve the lead contamination problem. But it will also presumably depend on the development of effective solutions to other problems considered in this book. Success in such efforts will surely demand a continued commitment toward conservation of the species by the public and on well-conceived research and management programs to overcome resistance factors.

The re-creation of a viable population of condors in the Northwest would constitute an achievement of substantial importance, not just for those with a special interest in birds, but perhaps especially for the many Native Americans living in the region, whose cultural traditions have always honored this species as a supreme master of the skies.

Noel Snyder
Former US Fish and Wildlife Service biologist in charge of condor research

The Pacific Northwest

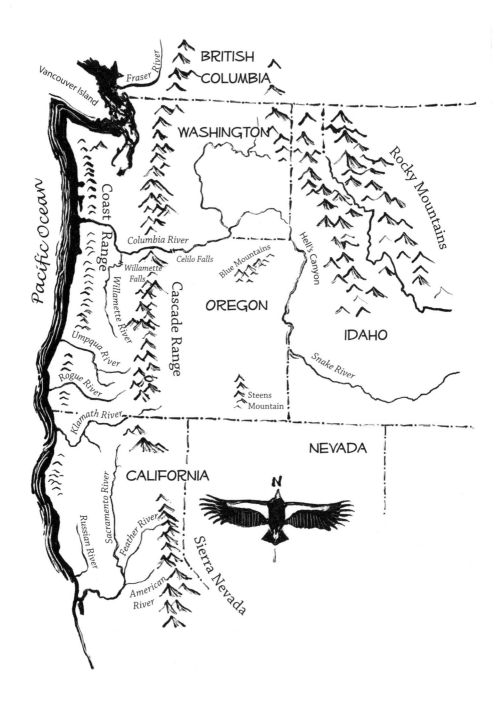

Preface

i•con (ī′kŏn′) *n.* An important and enduring symbol.

Fig. 1. Portrait of an endangered species icon: the California Condor. Photo by Michael Durham, Oregon Zoo.

The California Condor (*Gymnogyps californianus*), North America's largest avian scavenger and one of the largest flying birds in the world, is an iconic species by any measure (figure 1). Although commonly depicted as a bird of southern California and the desert Southwest, condors once soared the skies of the Pacific Northwest and were deeply woven into the fabric of many Native American cultures in the region. Described by Captain Meriwether Lewis as the "beatifull Buzzard of the columbia [river]," condors were observed and collected by members of the Lewis and Clark Expedition and other explorers, trappers, fur traders, naturalists, and settlers in many parts of the Northwest during the nineteenth century. Soon after 1900, however, the condor disappeared from its northern haunts and its population and range continued to contract throughout the twentieth century until only a small remnant population remained in the mountains of southern California.

Despite the extensive volume of literature published on California Condors and the Herculean conservation struggle to bring the condor back from the brink of extinction (reviewed by Snyder and Snyder 2000), to date relatively little attention has been paid to the history of condors in the Pacific Northwest and opportunities for restoring them to the region (but see Koford 1953; S. Wilbur 1973; Moen 2008; Sharp 2012). With the acceptance of the Oregon Zoo into the California Condor recovery program in 2001, and increasing interest in restoration of condors from a number of Native American tribes and the general public throughout the Northwest (see chapter 1), the need for a thorough review of the condor's history in the region has been building.

Understanding the history of condors in the northern half of their historical range is more than a curiosity. It is vital to the US Fish and Wildlife Service in defining recovery objectives and is a first step toward evaluating the potential for future reintroductions to the region. In its most basic sense, the history of a species' distribution and range collapse establishes context and helps one gauge the magnitude of anthropogenic changes over the last several hundred years rather than shifting the species' baseline condition to the current crisis situation (see Pauly 1995). It may also provide basic life history information of the species across its former range (e.g., historical breeding sites and movement patterns) that is important in setting appropriate recovery objectives. Finally, a species' natural history provides insights into the timing, magnitude, and causes of range collapse or population decline—information that is fundamental to assessing the restoration potential of imperiled species.

In this book, we document the California Condor's history in the Pacific Northwest through a review of anthropological, archaeological, paleobiological, and other historical information from myriad sources. We consulted published literature, unpublished reports, museum records, historical photographs, newspaper archives, early American journals, and documents at museums and state and federal resource management agencies, including the US Fish and Wildlife Service California Condor Recovery Office in Ventura, California, and the Santa Barbara Museum of Natural History. Our primary goal in writing this book is to provide an integrated and comprehensive synthesis of the condor's history in the region. However, it is also our hope that this book informs future dialogue concerning the role the Pacific Northwest might play in the recovery of this iconic species.

This book would not have been possible without the generosity and insights of many people. First and foremost, we thank Sanford Wilbur, principal condor researcher from 1970 to 1980, for his role as a key source of information on the historical occurrence records of California Condors in the region, something he has been investigating for decades. He also provided many constructive comments on draft chapters of this book. Sandy's continued work toward understanding the history of condors throughout their former range and sharing his knowledge with others is a testament to his undying commitment to this species.

We are indebted to Noel Snyder, principal condor researcher in the 1980s, for providing the foreword to the book and for suggesting several important corrections and additions that made the book more accurate and complete.

We also give special thanks to Jan Hamber. Jan's vast firsthand knowledge of the history of the condor recovery program and the extensive files she keeps at the Santa Barbara Museum of Natural History were most helpful. We also thank Jesse Grantham, California Condor Recovery Coordinator (2004–2012), for stimulating discussions on condor ecology and hosting us on a number of occasions to view condors in the wild and to sift through the official condor recovery files housed at the US Fish and Wildlife Service's Condor Recovery Office in Ventura, California.

When we could not find a copy of a report, book, or journal article, David Liberty, librarian at the StreamNet Regional Library in Portland, Oregon, was always willing to offer help. We are indebted to him for acquiring hard-to-find books and documents. We also thank the US Fish and Wildlife Service National Conservation Center Library and the Oregon Historical Society for their assistance in locating important documents.

Numerous archaeologists, paleontologists, anthropologists, and museum curators provided photographs and information related to condor bones and the use of condor parts in Native American cultures. Specifically, we thank Pamela Endzweig, Amanda Kohn, Martina Steffen, Patricia Nietfeld, Alison Stenger, and Jacob Fisher. We also thank Rich Young of the US Fish and Wildlife Service for his help in digitizing historical tribal boundaries, and Pepper Trail of the US Fish and Wildlife Service Forensics Lab for help in identifying feathers in historical photographs.

Many other people deserve recognition for sharing their knowledge of California Condors and guiding us to see condors in the wild. For this we thank Joseph Brandt, Joe Burnett, Eddie Feltes, Daniel George, Matthew Johnson, Chris Parish, Scott Scherbinski, Kelly Sorensen, and Mike Tyner.

We also thank Oregon Zoo personnel—Jane Hartline, David Shepherdson, David Moen, Kelli Walker, Michael Durham, Anne Warner, Shawn St. Michael, and Tony Vecchio—who provided information on the history of the zoo's entrée into the recovery program, as well as photos and access to the zoo's captive breeding facility.

We further thank Rolf Koford, Clint Epps, Bruce Marcot, David Shepherdson, Carrie Phillips, and Bruce Dugger for their helpful comments on earlier drafts of this book. Finally, we thank Ram Papish for providing texture to the book with his wonderful California Condor drawings. Support for our research was provided by the US Fish and Wildlife Service and US Geological Survey Forest and Rangeland Ecosystem Science Center. This book has been peer reviewed and approved for publication consistent with US Geological Survey Fundamental Science Practices (http://pubs.usgs.gov.circ/1367). The findings and conclusions are those of the authors and do not necessarily represent the views of the US Fish and Wildlife Service. Sections authored entirely by non-USGS authors do not represent the views or position of the US Geological Survey. Any use of trade or firm names is for descriptive purposes only and does not imply endorsement by the US government.

CALIFORNIA CONDORS

IN THE PACIFIC NORTHWEST

Chapter 1
Background

con•dor (kŏn'dôr'-dər) *n.* [Sp. *condor* < Quechua *kúntur.*] A very large New World vulture, *Vultur gryphus* of the Andes or *Gymnogyps californianus* of western North America.[1]

Evolution and Life History of the California Condor

Condors are often defined by their remarkable size. They are the largest of the seven New World vultures that form the Cathartidae family (sometimes referred to as the Vulturidae family; Livezey and Zusi 2007). Although New World vultures look similar to Old World vultures (Accipitridae family), this resemblance is the result of convergent evolution rather than a close phylogenetic relationship (Seibold and Helbig 1995; Wink 1995; Hackett et al. 2008).

California Condors have a truly spectacular wingspan (2.74 m)—larger than any other North American land bird. This large wingspan gives them the ability to soar long distances in a single day (at up to 40–70 km per hour), expending minimal energy while searching for food along Pacific Ocean beaches (figure 2) or inland over rivers, grasslands, and shrublands. However, their wingspan—and more specifically, the large surface area of their wings and weak wing musculature—limits their ability to sustain flapping flight for extended periods of time, as the immense amount of energy required to displace such a large quantity of air quickly exceeds their metabolic output (H. Fisher 1946; Pennycuick 1969). This means that they are restricted to foraging over areas where there is enough upward air movement, or lift, to keep them aloft. Such upward air movement is typically generated by thermals, which form when the sun heats the ground

---

1 Definitions are adapted from Webster's II New Riverside University Dictionary (1984) and have been abbreviated for clarity.

Fig. 2. California Condor soaring along the Big Sur coast, California. Photo by Jesse D'Elia, US Fish and Wildlife Service.

and the heated area causes a pocket of warm air to rise, or through ridge lift, whereby air is pushed upward as winds collide with mountains or cliffs.

Condor movements are influenced by the location of nests and foraging habitat. Breeding birds are necessarily tied to nest sites but may travel up to 180 km from the nest in search of food. Nonbreeding birds can move over enormous home ranges. For example, Meretsky and Snyder (1992) reported home ranges in southern California averaging approximately 7,000 km². California Condors do not undertake long-distance migrations but sometimes exhibit shorter seasonal movements to exploit traditional food resources or favorable atmospheric conditions.

Condors are obligate scavengers,[2] feeding primarily on medium to large-sized mammal carcasses, often including those of domestic livestock as well as native terrestrial and stranded marine mammals. Obligate scavengers are an example of extreme specialization in the animal kingdom, with several adaptations critical for species that rely on finding carrion, a relatively unpredictable and highly transient resource:

---

2   However, there was a recent case of a condor preying on an abandoned dying sea lion pup, and it is likely that condors occasionally take advantage of similar situations elsewhere (M. Tyner, Ventana Wildlife Society, pers. comm., 2011). Historical observations also suggest that condors once fed upon dead and dying salmon that were stranded as they attempted to move upstream toward spawning grounds (Audubon 1840). McGahan (2012) indicated that Andean Condors will apparently kill prey on rare occasions.

1. large size and large crops, which are necessary to compete at carcasses and sustain individuals for relatively long periods between meals;
2. soaring flight and excellent eyesight, which help condors efficiently find food;
3. hooked bills, long necks, and largely naked heads, which allow condors to access muscle tissue deep within a carcass and to rip pieces of meat from a carcass, while minimizing the potential for feather fouling;
4. feet with short claws adapted for walking and running, which may provide a competitive advantage at carcasses;
5. intelligence, which is necessary for finding and competing for food in a complex social environment; and
6. resistance to bacterial toxins, which is necessary for species that rely on carcasses.

Aerial scavengers can outcompete terrestrial scavengers for food because flight allows them to efficiently search a much larger area (Ruxton and Houston 2004). Soaring scavengers like the condor have an additional advantage over those that use primarily flapping flight because the energetic cost of soaring is so much less than that of flapping flight (Ruxton and Houston 2004).

Condors do not have a well-developed olfactory tract or sense of smell (Stager 1967), so they rely on their keen vision to find food. In addition to locating food from a distance, condors will also use sentinel species, such as Turkey Vultures (*Cathartes aura*) or Common Ravens (*Corvus corax*), to help them find food. This communal searching for food allows condors and other avian scavengers to greatly improve the efficiency with which they find a meal (Houston 1985, 1988). It also means that condors can be found congregated in large groups at a carcass or water hole. Condors are typically the dominant avian species at a carcass unless Golden Eagles (*Aquila chrysaetos*) are present.[3] As a social foraging species, they develop a pecking order at carcasses, with juveniles subordinate to adults. This may be a mechanism to reduce intraspecific aggression, but it also means that juveniles depend on their parents for an extended period to obtain sufficient food.

---

3 The Golden Eagle's dominance at carcasses is not absolute. Condors will occasionally challenge Golden Eagles and aggressively displace them from carcasses.

Condors' massive wingspans dictate their need for open spaces, good winds, and high places from which to launch flights. These factors are particularly important in selecting nest sites, where fledglings must learn how to fly. Nests are generally placed on the floor of small caves on cliff faces, on rock ledges, or occasionally in a cavity or broken top of a large tree. Breeding pairs mate for life and are intensely devoted to the care of the single egg they lay and the resulting chick. Maximum productivity for a pair appears to be two surviving chicks in three years, as clutch size is always one, full nesting cycles take more than a year, and pairs are slow to reinitiate breeding when they still have a dependent fledgling (Meretsky et al. 2000). Condors will double clutch if their egg is removed or destroyed (Snyder and Hamber 1985).[4]

California Condors are long-lived (Mace 2011).[5] Yet population growth rates are slow, as birds generally do not successfully breed until they are six to eight years old. Their slow maturation, slow breeding cycle, and low fecundity make condor populations sensitive to increases in adult mortality (Meretsky et al. 2000). These factors also make recovery of the species a relatively long and expensive proposition.

Readers looking for a more thorough review of the evolution and life history of the California Condor are directed to Koford (1953), Snyder and Snyder (2000, 2005), and Snyder and Schmitt (2002). For an overview of the evolutionary adaptations of scavengers, see Houston (1979, 1985, and 1988) and Ruxton and Houston (2004).

A Brief History of the Condor Recovery Program

The California Condor Recovery Program is one of the oldest and most renowned recovery efforts in the history of endangered species conservation. It is also one of the most controversial. Below we give a brief overview of the Condor Recovery Program, beginning with the development of the

---

4 Although *apparent* double clutching in condors was first reported by Harrison and Kiff in 1980, their paper showed only that two nesting events took place in the same year in the same cave. Because the adults were never identified, evidence from this paper was insufficient to get permission to take eggs from nests to start a captive flock. It was not until 1982 that conclusive evidence of replacement clutching was obtained (Snyder and Hamber 1985), allowing the start of egg removal operations to form a captive population the next year.

5 Topa-Topa (studbook #1), a male condor and the oldest condor in captivity, hatched in the wild in Ventura County in 1966. He was captured in 1967 and, as of 2012, was still alive at the Los Angeles Zoo (Mace 2011).

first US Fish and Wildlife Service recovery plan in 1975. Additional details are given in table 1, which provides a concise timeline of important events in the history of the California Condor and its recovery. Readers seeking a more thorough treatment of the history of the recovery program and the early field studies of California Condor natural history are directed to Koford (1953), McMillan (1968), S. Wilbur (1978), Snyder and Snyder (2000), S. Wilbur (2004), Alagona (2004), and Walters et al. (2010).

In retrospect, it is remarkable that the California Condor did not end up as yet another entry in the long ledger of extinct birds (see Fuller [2001] for a detailed accounting of those birds that have been lost). At the time the first recovery plan for the California Condor was published in 1975—the first recovery plan for any species under the US Endangered Species Act of 1973— there were only about forty condors remaining (S. Wilbur 1978). By 1980 that estimate was reduced to twenty-five to thirty-five individuals (S. Wilbur 1980). Clearly the species was in jeopardy of going extinct if something was not done quickly to reverse the decline. The plan's population objectives were modest, calling for the maintenance of "at least 50 California Condors, well distributed throughout their 1974 range" (USFWS 1975, 12). The recovery team noted the possibility that recovery efforts in the wild might fail and suggested developing a contingency plan that included captive breeding.

> While the condors' inherently low reproductive capacity makes it appear a less likely candidate for captive propagation than some other species, recent successes at the Patuxent Wildlife Research Center and elsewhere propagating South American condors (*Vultur gryphus*) gives some hope for the future of this technique. Patuxent personnel plan a continuing investigation of South American condor propagation and subsequent release to the wild, and this may have application to the California condor should current plans fail to improve its population status. (USFWS 1975, 11)

In 1976, the US Fish and Wildlife Service California Condor Recovery Team, faced with continuing declines in condor population estimates and the very real possibility that the condor was vanishing, drafted a contingency plan that included provisions for initiating a captive-breeding program, with a view toward future reintroductions, a suggestion that was

Table 1. Timeline of events in the history and recovery of the California Condor.

Year/ time period	Event	No. of wild California Condors*
Pleisto-cene	Several species of *Gymnogyps* occur in North America. The California Condor is distributed from British Columbia to Baja California, and inland to Arizona, New Mexico, Texas, and central Mexico. It also occurs along the east coast from New York to Florida.	Unknown
Late Pleisto-cene	All species of *Gymnogyps* aside from the California Condor go extinct. The California Condor's range contracts and it is now limited to the West Coast of North America from British Columbia to Baja California. Paleo-Indians begin populating North America.	Unknown
1602	First recorded sighting of a California Condor by European explorers—Father Antonio de la Ascension in Monterey Bay, California.	Unknown
1790s	Type specimen taken near Monterey, California, by Archibald Menzies. This condor skin is now housed at the British Natural History Museum at Tring.	Unknown
1805	Lewis and Clark and the Corps of Discovery observe California Condors along the lower Columbia River from Celilo Falls to the coast.	Unknown
1849	California Gold Rush—massive influx of people to northern California.	Unknown
1850	California Condors no longer regularly reported from the lower Columbia River. Still sporadically collected and reported from elsewhere in the Pacific Northwest.	Unknown
1904	Generally regarded as the last reliable report of condors north of San Francisco, CA. However, there are a few other plausible reports of condors in the region into the 1920s.	Unknown
1905	Killing or collecting condors or their eggs is banned by the California Legislature and Fish and Game Commission.	Unknown
1906	William L. Finley and Herman T. Bohlman make the first study of a condor nest. Finley takes the chick captive and raises him as a pet in Oregon before transferring him to the New York Zoological Park.	Unknown
1939	Carl Koford begins his landmark study of the California Condor.	150
1940s	San Diego Zoo is breeding Andean Condors successfully and demonstrates that pairs can produce more than one egg a year through replacement clutching.	150
1949	Belle Beachy of the San Diego Zoo proposes captive breeding of California Condors to the California Department of Fish and Game. Although the department approves the zoo's proposal to capture two immature condors, trappers fail to catch any birds.	150
1953	Carl Koford completes the first major natural history study of the California Condor (Koford 1953).	150
1954	California Legislature expressly forbids taking any California Condors from the wild. San Diego Zoo trapping efforts cease.	150
1966	The US Congress passes the Endangered Species Preservation Act on 15 October 1966.	60

Year/ time period	Event	No. of wild California Condors*
1967	California Condor designated an endangered species under the Endangered Species Preservation Act.	60
1969	Locke et al. (1969) discover that Andean Condors are susceptible to lead poisoning and suggest that California Condors might also be susceptible.	60
1973	Endangered Species Act (ESA) of 1973 passed and additional protections given to species listed in 1967 under the Endangered Species Preservation Act.	35–60
1975	The California Condor Recovery Team is established and the Condor Recovery Plan is adopted. It is the first recovery plan for any species under the US Endangered Species Act.	25–35
1976	Designation of California Condor critical habitat under the Endangered Species Act—all in southern California.	25–35
1980	First revision to the California Condor Recovery Plan adopted. Recommends captive breeding and identification of release sites by surveying areas of former occupation—including areas in the Pacific Northwest.	25–35
1982	Nadir of the California Condor population (considering both captive and wild birds, only 22 remain).	20
1983	Taking eggs from wild nests for artificial incubation and captive rearing initiated.	< 20
1984	Second revision to the California Condor Recovery Plan adopted. The primary objective of the plan is to increase and maintain a self-sustaining population of 100 individuals, including 60 adults. Recommends captive breeding and multiple clutching of wild nesting pairs.	< 20
1985	Catastrophic loss of 40 percent of the remaining wild condors (cause of death unknown). Discussions initiated regarding trapping all remaining wild condors.	< 10
1987	Last wild California Condor (AC-9) trapped for captive breeding. The California Condor is extinct in the wild. At this time, 27 condors are in captivity (10 reared in the wild, 17 reared in captivity).	0
1988	Experimental releases of Andean Condors into southern California initiated.	0
1991	California Condor Recovery Team recommends releases in northern Arizona in addition to releases in southern California.	0
1992	Releases of Andean and California Condors on the Sespe Condor Sanctuary. Congress passes an appropriations rider granting federal money to the Peregrine Fund to breed condors and release them near the Grand Canyon, Arizona.	7
1993	Third captive breeding facility established—World Center for Birds of Prey in Boise, Idaho, operated by the Peregrine Fund.	9

Year/ time period	Event	No. of wild California Condors*
1994	California Condors retrapped due to behavioral problems.	3
1995	Release of California Condors that had undergone aversion training to reduce behavioral issues. The three condors that remained in the wild in 1994 were trapped to ensure they did not negatively influence the newly released birds that underwent aversion training.	14
1996	Second revision to the California Condor Recovery Plan adopted. Drops mention of identifying release sites in the Pacific Northwest. Focuses on building population levels to at least 150 birds in southern California and 150 birds in Arizona. Does not identify actions needed to achieve recovery; only identifies downlisting to threatened status criteria under the ESA. The USFWS publishes a final experimental population rule designating northern Arizona, southern Utah, and a small corner of southeastern Nevada as a "non-essential experimental population." Condor releases begin in December 1996 at Vermilion Cliffs, northern Arizona.	17
1997	Condor releases begin near Big Sur, Monterey County, California.	29
1998	Condor releases begin at Hurricane Cliffs in northwestern Arizona, 65 miles west of the Vermilion Cliffs release site (later discontinued due to logistical issues).	38
2001	Oregon Zoo presents its proposal to breed condors to the Condor Recovery Team with the ultimate goal of reintroducing them to Oregon. The proposal is accepted by the Recovery Team.	58
2002	Condor releases begin in Baja California, Mexico.	71
2003	Condor releases begin at Pinnacles National Monument, California. First captive condors arrive at the Oregon Zoo's Jonsson Center for Wildlife Conservation.	83
2004	First California Condor egg hatched at the Oregon Zoo's Jonsson Center for Wildlife Conservation.	96
2007	The Yurok Tribal Council passes a resolution to develop a California Condor reintroduction site.	144
2008	The US Fish and Wildlife Service provides funds to the Yurok Tribe to study the feasibility of reintroducing California Condors to northern California.	167
2010	California Condor review panel commissioned by the American Ornithologists' Union and the Audubon Society publishes its review of the recovery program (Walters et al. 2010).	181
2011	First meeting of the Pacific Northwest California Condor Coordination Team, an interdisciplinary and interagency team organized by the USFWS to evaluate remaining issues that need to be resolved prior to establishing a Pacific Northwest condor release site.	205

* Population estimates through the 1980s are based on Snyder and Snyder (2000). The numbers of wild condors from the 1990s through 2011 are based on California Condor Recovery Program records.

later supported by a panel of ornithologists and the National Audubon Society (Ricklefs 1978), and by a report prepared for the US Forest Service (Verner 1978). On 2 November 1978, the director of the US Fish and Wildlife Service met with representatives of the National Audubon Society, who presented their recommendations for modifying the condor recovery strategy (USFWS 1979). This meeting resulted in the formation of a task force charged with charting a course for implementing a captive breeding program and identifying areas appropriate for future releases (USFWS 1979). Consequently, when the first revision to the condor recovery plan was published in 1980, it included the need to initiate captive breeding and identify potential reintroduction sites "in the states occupied by condors in the recent past (Oregon, Washington, California, possibly Arizona)" (USFWS 1980, 50).

Although captive breeding was initially meant to supplement the wild population (USFWS 1984), a catastrophic loss of 40 percent of the remaining population in the winter of 1984–1985 left only a single breeding pair in the wild. This led the recovery team and partners to reevaluate whether the wild population should be supplemented or integrated with the captive population to maximize genetic diversity (Snyder and Snyder 2000). Geneticists advising the Condor Recovery Team agreed that because of the limited number of individuals and family lines remaining, all remaining wild birds should be immediately added to the captive flock (Snyder and Snyder 2000). However, disagreements among the US Fish and Wildlife Service, the Audubon Society, the Condor Recovery Team, and the California Fish and Game Commission—and ultimately, litigation brought by the Audubon Society—delayed that decision (Snyder and Snyder 2000).[6]

Although capturing all remaining wild condors to form a captive breeding population was extremely controversial (L. Miller 1953; Pitelka 1981;

---

6 The US Fish and Wildlife Service, at the urging of the Audubon Society, did not support trapping all remaining birds at the beginning of 1985; instead, it advocated trapping only three birds and simultaneously releasing three captive birds. At the time, the USFWS apparently thought that mortality risks for the remaining birds could be significantly reduced through an intensive food provisioning program with lead-free carcasses (Snyder and Snyder 2000). Lead poisoning of a condor (AC-3) in December of 1985 on Hudson Ranch— where lead-free carcasses were being provided—ended the debate over the efficacy of food provisioning in reducing mortalities (Snyder and Snyder 2000). Litigation by the Audubon Society delayed the final trapping of all remaining birds until the spring of 1986.

Snyder and Snyder 2000; Alagona 2004), it was the only hope of preserving the species (Snyder and Snyder 2000). At the time the last wild condor was trapped, on Easter Sunday 1987, only twenty-seven California Condors remained in the world.

With all California Condors in captivity there was an urgent need to work out the most effective methods for minimizing mortality in future releases (Wallace 1989). Fortunately, Mike Wallace, curator of birds at the Los Angeles Zoo and a member of the California Condor Recovery Team, had completed a dissertation based on captive releases of Andean Condors (*Vultur gryphus*) in Peru from 1980 to 1984. Beginning in 1988, Wallace assisted with experimental releases of Andean Condors in southern California as surrogates for future California Condor releases.

Captive breeding and double-clutching protocols for California Condors were established by the early 1990s (Meretsky et al. 2000). In December 1991, the Condor Recovery Team recommended that releases also be conducted in northern Arizona in an area geographically separate from the southern California flock (USFWS 1996a).

Surveys of suitable habitat were never conducted in Oregon, Washington, or northern California, and subsequent revisions to the recovery plan dropped any mention of reintroductions to the Pacific Northwest (USFWS 1984, 1996b). Although significant progress had been made in captive breeding and release techniques through the early 1990s, there were only seventeen condors in the wild at the time of the last recovery plan revision in 1996. Furthermore, the plan did not address full recovery. Instead, its emphasis was on how to improve the species' status to the point where it could be reclassified from endangered to threatened under the Endangered Species Act by maintaining a captive flock and establishing self-sustaining populations in southern California and Arizona (USFWS 1996b). Specifically, the recovery criteria for reclassification from endangered to threatened read:

> The minimum criterion for reclassification to threatened is the maintenance of at least two non-captive populations and one captive population. These populations (1) must each number at least 150 individuals, (2) must each contain at least 15 breeding pairs and (3)

be reproductively self-sustaining[7] and have a positive rate of population growth. In addition, the non-captive populations (4) must be spatially disjunct and non-interacting, [and] (5) must contain individuals descended from each of the 14 founders. (USFWS 1996b)

Since the 1996 recovery plan, captive breeding efforts have proven extremely fruitful in boosting condor numbers. Since 1993, over three hundred condors have been raised in captivity and released into the wild and there are now five active release sites: Big Sur, California; southern California mountains; Pinnacles National Monument, California; northern Arizona; and Sierra San Pedro Mártir, Baja California, Mexico (Walters et al. 2010). With growing numbers of condors in captivity and in the wild, years of experience from several release programs, and a greater understanding of condor biology, population threats, and conservation needs, there is now growing interest by conservation organizations and Native American tribes in reestablishing the condor in the Pacific Northwest (Shepherdson et al. 2007; The Nature Conservancy, in litt. 2007; Yurok Tribe 2007; Walters et al. 2010).

Bringing Condors Back to the Pacific Northwest: The Birth of an Idea

As captive breeding techniques were worked out in the 1980s and 1990s and the captive population began to grow, the capacity of zoos in the program to breed and house condors became a limiting factor. Consequently, in the late 1990s and early 2000s, the Condor Recovery Team began looking for additional zoos that were interested in joining the conservation breeding program. Adding another conservation breeding partner would have the benefits of spreading the risk to the captive population (e.g., containing a disease outbreak to only a portion of the population), increasing capacity to produce condors, and sharing the substantial costs associated with captive breeding and rearing. Several zoos expressed interest in joining the recovery effort, including the Bronx Zoo, the National Zoo, and the Oregon Zoo.

The Oregon Zoo's interest in condor conservation and the notion of California Condor reintroductions to the Pacific Northwest stemmed from

---

7 Although the recovery plan used the term "self-sustaining," it also recognized (and allowed) that in some areas, reestablished condor populations might require continued artificial feeding to supplement natural food resources and/or to protect birds from exposure to contaminated carcasses.

planning sessions for the Lewis and Clark bicentennial (Koch 2004). In preparation for the bicentennial, Jane Hartline, then marketing manager for the zoo, suggested reintroducing California Condors to Oregon. Jane's idea was sparked by her recent trip to Ecuador, where she visited Hacienda Zuleta, a hotel on a colonial working farm that hosts an Andean Condor rehabilitation and educational facility.

As an outgrowth of the bicentennial planning sessions, the Oregon Zoo initiated discussions with the Condor Recovery Team in 2000 and presented a proposal to join the recovery program as a captive breeding facility in February 2001, with hopes of eventually reintroducing condors to Oregon (Koch 2004). After considering proposals from a number of zoos, the Condor Recovery Team accepted the Oregon Zoo's proposal later that year.

Upon acceptance into the recovery program, the zoo immediately began the process of selecting an offsite location for breeding condors. Because condors develop behavioral problems when they have contact with humans, the zoo sought a property that was out of view of the public. This ultimately led to the construction of a state-of-the-art condor breeding and veterinary facility at the Jonsson Center for Wildlife Conservation, in an undisclosed rural location near Portland, Oregon (figure 3). The first pair of condors arrived at the facility on 19 November 2003, and the first egg hatched there the following spring (Koch 2004). It is now one of four facilities that breed condors for release into the wild.[8]

The Oregon Zoo is not the only organization interested in returning condors to the northern portion of their historical range. In 2007, the US Fish and Wildlife Service (USFWS) received a grant proposal from the Yurok Tribe in northwestern California to assess the feasibility of reintroducing condors to their ancestral lands (Yurok Tribe 2007). The Yurok believe that reintroduction of condors to the tribe's ancestral territory will help restore spiritual balance to their world (Yurok Tribe 2007). The Yurok Tribe and the Oregon Zoo formed an informal partnership to promote the idea of reintroductions and in April 2010, the tribe and the zoo hosted a Pacific Northwest Condor Summit (with sponsorship from the Confederated Tribes of Grand Ronde), bringing together over 140 participants from other

---

8   The others are the San Diego Wild Animal Park, Los Angeles Zoo, and the World Center for Birds of Prey, the latter operated by the Peregrine Fund in Boise, Idaho.

Fig. 3. California Condor at the Oregon Zoo's Jonsson Center for Wildlife Conservation, just outside of Portland, Oregon. Photo by Susan Haig, US Geological Survey.

Northwest tribes, federal and state agencies, and conservation groups, as well as representatives from the California Condor Recovery Program.

With growing interest in returning condors to the Pacific Northwest, a number of questions remain: What was the historical distribution of condors in the region? Were they breeding here, or simply seasonal migrants? When did they disappear? What caused their extirpation? What is the potential for restoration (i.e., have the primary threats been identified and ameliorated and would ongoing management be necessary)? Where are the best places for a reintroduction in the Pacific Northwest? And, are there lessons that might be learned from other vulture reintroduction projects throughout the world? In the next few chapters we hope to provide answers to some of these questions.

Chapter 2
Historical Distribution of California Condors: A Review of the Evidence

dis•tri•bu•tion (dĭs'trə-byōō'shən) n. 1. The act of distributing or the condition of being distributed. 2. Bio. The geographic occurrence or range of an organism.

Prehistoric Distribution and Pleistocene Range Contraction

Fossil evidence suggests that California Condors were widely distributed in North America during the late Pleistocene,[1] with records from Oregon, California, Nevada, Arizona, New Mexico, Texas, Florida, New York, and Mexico (L. H. Miller 1910a, 1910b, 1911; L. Miller 1957; Wetmore 1931a, 1931b; Parmalee 1969; Simons 1983; Steadman and Miller 1987; Emslie 1987; Steadman et al. 1994; Hansel-Kuehn 2003; Brasso and Emslie 2006). At that time, megaherbivores[2] and megapredators[3] likely provided significant food resources for large avian scavengers—including several condor species (*Gymnogyps californianus, G. amplus, G. varonai, G. kofordi, Breagyps clarki*)—that relied on the availability of large carcasses for survival (Emslie 1987, 1990).

The leading hypothesis regarding the California Condor's range contraction and the extinction of all of its congeners at the end of the Pleistocene is the loss of sufficient food resources away from coastal areas (Emslie 1987, 1990; Steadman and Miller 1987; Suárez 2000; Chamberlain et al.

---

1 circa 50,000–10,000 years before present (YBP)

2 e.g., ground sloths (*Megalonyx jeffersonii, Eremotherium laurillardi, Nothrotheriops shastensis, Glossotherium harlani*), camels (*Camelops* spp.), wild horses (*Equus* spp.), giant bison (*Bison latifrons*), shrub oxen (*Euceratherium* spp.), mastodons (*Mammut americanum*), and woolly mammoths (*Mammuthus primigenius*)

3 e.g., dire wolves (*Canis dirus*), saber-toothed tigers (*Smilodon* spp.), and short-faced bears (*Arctodus simus*)

2005; Fox-Dobbs et al. 2006). This hypothesis is consistent with available, albeit limited, radiocarbon-dated fossil condor evidence and stable isotope composition of bone collagen from museum specimens (Emslie 1987, 1990; Fox-Dobbs et al. 2006). Although prehistoric condors in Florida and Texas presumably had access to marine food resources, the lack of large salmon runs, pinniped rookeries, and haul-out sites, as well as the lack of offshore coastal upwelling (the wind-driven oceanographic phenomenon where cooler nutrient-rich water replaces warmer surface water, leading to increased productivity and food availability for large marine mammals), would have meant that food resources for scavengers may have become more limited there than along the Pacific coast in the wake of the terrestrial megafauna extinctions (Fox-Dobbs et al. 2006). However, other synergistic factors, including rapid climatic cooling and drought during the Younger Dryas (12,800 and 11,500 YBP)[4], changes in the availability of thermal updrafts for soaring and finding food, and increased competition from other scavengers or mesopredators, may have also played a role and have not been sufficiently explored.

Along the Pacific coast there is no evidence that food was limiting for condors during the late Pleistocene or during the Holocene. Although the largest land mammals went extinct at the end of the Pleistocene, virtually all the surviving large herbivorous mammals (e.g., deer, elk) are wide-ranging species (Martin 1990) and were likely abundant in the Pacific Northwest, especially west of the Cascades and Sierra Nevada crest, where temperate forests looked much as they do today (Hansen 1947). Furthermore, beached whales, abundant salmon runs, and numerous pinniped rookeries and haul-out sites were also available to condors along the coast. In addition to food availability, the topographic complexity of the western

---

4 The cause of rapid cooling and drought during the Younger Dryas has been the subject of significant debate (see Broecker et al. 2010). The most pervasive hypothesis proposes that a catastrophic release of fresh water from proglacial Lake Agassiz shut down the Atlantic Ocean's circulation, resulting in an extensive winter sea ice cover whose presence blocked the release of ocean heat, directed westerly winds to the south, and reflected solar radiation. More recently, largely as a result of dramatic recreations on television, the hypothesis that the Younger Dryas was triggered by the impact of an asteroid (or multiple asteroids, or airbursts over the Laurentide Ice Sheet) has gained public attention, but it has not gained traction in the scientific community. Broecker et al. (2010) argued that the Younger Dryas appears to be simply a stall in the processes that brought the last glacial period to a close and that no single catastrophic event is required to explain its existence.

United States provides extensive areas of slope lift that allow soaring avian scavengers to search vast areas for food, even in the absence of strong year-round thermals. Finally, the warming influence of the Pacific Ocean meant that climatic changes along the Pacific Northwest coast during the Younger Dryas were less dramatic than changes experienced in the Intermountain West and Great Plains (McCornack 1920; Hansen 1947). Whatever the reason, condors persisted in the Pacific Northwest for many millennia, alongside Paleo-Indians and Native Americans, prior to the arrival of early Euro-American and Russian explorers.

The Archaeological and Paleontological Record

Bones of two condor species—*Gymnogyps californianus* and *G. amplus*—have been found at fifteen prehistoric sites in the Pacific Northwest, ranging in age from about 200 to 25,000 YBP (table 2, figure 4). A larger condor-like species (*Teratornis woodburnensis*) with a 4.25 m wingspan was also present in the Pacific Northwest during the late Pleistocene (approximately 12,000 YBP), based on bones (humerus, parts of the cranium, beak, sternum, and vertebrae) unearthed from a buried bog in Woodburn, Oregon (northern Willamette Valley; Campbell and Stenger 2002; figure 4).[5]

Sites containing *Gymnogyps* fossils are distributed from San Francisco to southern British Columbia and from the Pacific Ocean beaches inland to the western slopes of the Sierra Nevada in California and along the Columbia River near The Dalles, Oregon (immediately east of the Cascade Range; figure 4). The recent discovery of a broken California Condor tarsometatarsus[6] on South Pender Island, British Columbia (I. R. Wilson Consultants, Ltd. 2006; figure 5), extends the northernmost paleontological record of condors by approximately 400 km. Although there was a report of bones possibly belonging to the California Condor at Smith Creek Cave in northeast Nevada (Howard 1952), paleontologist Steven D. Emslie later deter-

---

5 Although Teratorns had large wingspans and several characteristics in common with Cathartid vultures (L. Miller 1909), more recent analyses of cranial morphology suggest that they were predators (likely piscavores) (Campbell and Tonni 1981, 1983; Hertel 1995). However, the abundant remains of Teratorns at the Rancho La Brea Tar Pits suggest that they may have also been facultative scavengers, similar to Bald Eagles (*Haliaeetus leucocephalus*; Hertel 1995).

6 A large bone in the lower leg of birds with which the toes connect, formed by fusion of the tarsal and metatarsal bones.

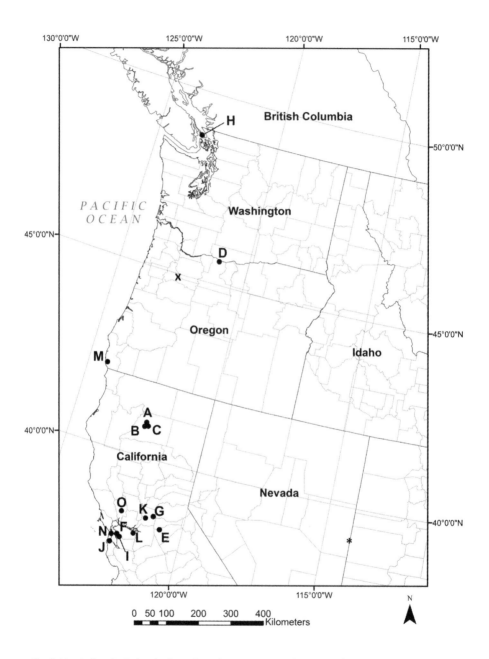

Fig. 4. (•) = Archaeological and paleontological sites containing *Gymnogyps* bones (see table 2 for additional details); (x) = Woodburn archaeological site, where bones of the extinct condor-like *Teratornis woodburnensis* were found; (*) = Smith Creek Cave, where remains of extinct large avian scavengers (*Aiolornis incredibilis* [formerly *Teratornis incredibilis*], *Breagyps clarki*, and *Coragyps occidentalis*) were found.

Table 2. The fossil record of *Gymnogyps* in the Pacific Northwest.[a]

Site letter	Site name	Location	Minimum number of bones	Species	Site type[b]	Years before present[c]	References
A	Potter Creek Cave	Shasta County, CA	10	G. amplus	N	25,000–14,100	Sinclair 1904; L. Miller 1911; Payen and Taylor 1976; Feranec 2009
B	Samwel Cave	Shasta County, CA	2	G. amplus	N	21,010–15,980	L. Miller 1911; Feranec et al. 2007
C	Galen's Pit	Shasta County, CA	3	G. californianus?	N	20,970–20,350*	University of California Museum of Paleontology specimen 131552; Emslie 1990
D	Five Mile Rapids (ORE-WS-4)	Wasco County, OR	212	G. amplus?	C	11,970–6,750	L. Miller 1957; Simons 1983; Hansel-Kuehn 2003; Syverson and Prothero 2010
E	Old Bridge site (CA-Cal-237)	Calaveras County, CA	2 humeri wands	G. californianus	B	3,930–3,330	Johnson 1967; R. Taylor 1975; Simons 1983
F	West Berkeley shellmound (CA-Ala 307)	Alameda County, CA	1 articulated skeleton	G. californianus	B	3,510–2,400	Wallace and Lathrap 1959; Crane 1956; Wallace and Lathrap 1975; Simons 1983
G	Windmiller mound (CA-Sac-107c)	Sacramento County, CA	1 mandible	G. californianus	B	3,180–2,540	Gifford 1940; Ragir 1972; Simons 1983
H	Poet's Cove, Bedwell Harbour (DeRt-4)	South Pender Island, BC	1	G. californianus	C	2,935–2,905*	I. R. Wilson Consultants, Ltd. 2006; Royal British Columbia Museum Archaeology Collection
I	Emeryville shellmound (CA-Ala-309)	Alameda County, CA	partial skeleton	G. californianus	B	2,690–660	Uhle 1907; Howard 1929; Wallace and Lathrap 1959; Simons 1983; Broughton 2004

Site letter	Site name [a]	Location	Minimum number of bones	Species	Site type[b]	Years before present[c]	References
J	Stevenson Street (CA-SFR-112)	San Francisco County, CA	1	G. californianus	C	1,790–1,240	Lieberson 1988; Broughton et al. 2007
K	McCauley mound number 3 (CA-Sjo-43)	San Joaquin County, CA	4 bone whistles	G. californianus	C	1,500–500	Schenck and Dawson 1929; Simons 1983; Maniery 1991
L	Hotchkiss mound (CA-Cco-138)	Contra Costa County, CA	3 bone whistles	G. californianus	C	1,429–350	Libby 1954; Crane and Griffin 1960; Simons 1983
M	Brookings site (Lone Ranch Creek shellmound) (35CU27)	Curry County, OR	1	G. californianus	C	1,090–220	A. Miller 1942; Berreman 1944; Simons 1983; Moss and Erlandson 2008; M. Moss, University of Oregon Department of Anthropology, pers. comm., 2011
N	Filoli site (CA-SMA-125)	San Mateo County, CA	4 bone whistles	G. californianus	B	ca. 1,000	Morejohn and Galloway 1983; Griffin et al. 2006
O	Berryessa Valley Adobe	Napa County, CA	1 bone whistle	G. californianus	C	ca. 200	Simons 1983

a   See figure 4 for a map of the sites.
b   B = Burial; C = Cultural Deposit; N = Natural Deposit
c   Based on radiocarbon dates from associated material at the site.
*   Asterisk indicates direct radiocarbon date of a condor bone.

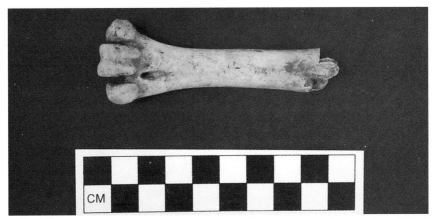

Fig. 5. Partial California Condor tarsometatarsus unearthed from Poets Cove archaeological site (DeRt-4), Bedwell Harbour, South Pender Island, British Columbia, Canada. Image B-01601 courtesy of the Royal British Columbia Museum, British Columbia Archives, Victoria, British Columbia, Canada. This bone was radiocarbon dated to 2,920±15 YBP (I. R. Wilson Consultants, Ltd. 2006).

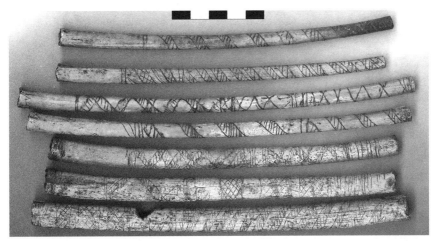

Fig. 6. Bird bone tubes and whistles from an archaeological site in Watsonville, California (CA-SCR-44). The bottom whistle was made from the left ulna of a California Condor (Breschini and Haversat 2000). The other bones are from Bald Eagles (*Haliaeetus leucocephalus*; the two above the condor bone) and Great Blue Herons (*Ardea herodias*; the four bones at the top). Photo by Trudy Haversat and Gary S. Breschini. Used with permission of Gary S. Breschini, Archaeological Consulting, Salinas, California. Scale in centimeters.

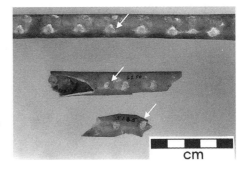

Fig. 7. Ancient Gymnogyps bones from the Five Mile Rapids archaeological dig near The Dalles, Oregon. White arrows show cut marks along the papillae that suggest feather removal, probably for ceremonial regalia. Photo by Henry Moore, Jr.

mined that those bones were from the extinct Clark's Condor (*Breagyps clarki*; Steadman and Miller 1987; Emslie 1988).[7] Most of the archaeological and paleontological evidence includes one or a few fragmentary bones (table 2). In one case, however, a nearly complete, articulated skeleton— including the entire skull—was found preserved at a Berkeley, California, shellmound, suggesting a ceremonial burial (Wallace and Lathrap 1959). At several other sites around San Francisco and Sacramento, condor whistles and ornamented bone wands were found in association with human burials (table 2; figure 6). Over two hundred condor bones were unearthed at the Five Mile Rapids archaeological site near The Dalles, with approximately 20 percent of the bones (humerus, radius, and ulna) containing cut marks, signifying disarticulation and feather removal (L. Miller 1957; Hansel-Kuehn 2003; figure 7), presumably for ceremonial regalia. At least one bone (the proximal end of a radius) also showed signs of being girdled (Hansel-Kuehn 2003), indicating that it may have been used as a primitive tool.

The archaeological and paleontological record establishes that condors occupied the region from the Cascade Range and Sierra Nevada to the coast for many millennia prior to Euro-American contact. Apart from the archaeological dig near The Dalles, evidence of ancient California Condor bones is currently lacking in the drier areas of the Pacific Northwest east of the Cascades and Sierra Nevada. However, additional evidence of condors from this area should be expected, as late Pleistocene California Condor bones have been found at dry inland sites to the south, and the Great Basin supported a diverse assemblage of large soaring scavengers and predators during this time period, including the Clark's Condor, the Western Black Vulture (*Coragyps occidentalis*), the Incredible Teratorn (*Aiolornis incredibilis*), and Merriam's Teratorn (*Teratornis merriami*) (Grayson 2011).

An evaluation of bones found in limestone caves in the lower McCloud River area near Mount Shasta in northern California indicates that a larger extinct condor (*Gymnogyps amplus*) was also once present in the Pacific Northwest (L. H. Miller 1911; table 2). Syverson and Prothero (2010), noting that the larger size of the single condor skull found at the Five Mile Rapids site suggests that this specimen also belonged to *G. amplus*, stated: "The skull of UCMVZ 1337 falls well outside the range of sizes observed

---

7 Although Howard (1952) tentatively identified six bones as *Gymnogyps*, she questioned this identification and suggested that they might be *Breagyps*.

for *G. californianus*, and indeed at the large end of the spectrum of RLB [Rancho La Brea] specimens" (11). Unfortunately, Syverson and Prothero included only a small sample of the bones from the Five Mile Rapids site in their analysis. While they had access to the twenty-one condor bones from that site that are retained at the Museum of Vertebrate Zoology in Berkeley, California (at the request of L. Miller in a letter of 7 July 1955 to L. Cressman, State Museum of Anthropology, University of Oregon, Eugene), the majority of bones from the Five Mile Rapids site are housed at the University of Oregon Museum of Natural and Cultural History (Hansel-Kuehn 2003) and were not included in their study.

Gymnogyps bones found at Galen's Pit, near Mount Shasta (table 2), were radiocarbon dated to 20,660±310 YBP and identified by Emslie (1990) as *G. californianus*. However, Emslie did not recognize *G. amplus* as a separate species (Emslie 1988), instead assigning the larger form as a possible temporal subspecies of *G. californianus*. As stated above, new information now suggests *G. amplus* may have been a distinct species (Syverson and Prothero 2010). Thus, reevaluation of Emslie's assignment of the Galen's Pit *Gymnogyps* bones is warranted.

Future work to reexamine the species-level identity of *Gymnogyps* bones from Galen's Pit and the Five Mile Rapids site in light of Syverson and Prothero's (2010) study may help clarify the evolutionary relationships and temporal co-occurrence of these two condor species in the Pacific Northwest. Extraction of ancient DNA from these bones may provide another complementary avenue of inquiry.[8] Whatever approach is taken, additional work in this area should be coupled with more direct radiocarbon dates from Pacific Northwest condor bones, as only two such dates are currently available (table 2).

California Condors in Pacific Northwest Native American Culture

Condors were known to many Pacific Northwest tribes, who idolized them in their mythology, depicted them in their art, and used their feathers in ceremonies, medicine, and dances. Below, we review the ethnographic record of tribal relationships with the condor in the Pacific Northwest by

---

8 The authors have taken a small sample from one of the condor bones from the Five Mile Rapids site and hope to extract ancient DNA to compare with contemporary California Condor DNA.

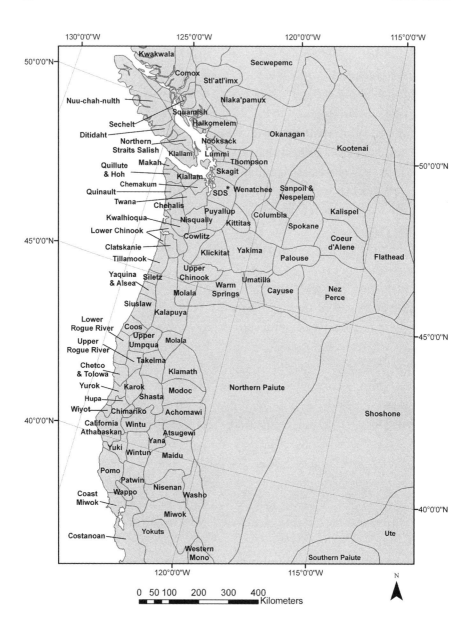

Fig. 8. Approximate tribal boundaries in the Pacific Northwest at the time of Euro-American contact. Boundaries and names based on Sturtevant (1967) and a map of historical First Nations languages of British Columbia developed by the University of British Columbia's Museum of Anthropology in coordination with the First Peoples' Heritage, Language and Culture Council. These boundaries were diffuse and dynamic but are shown here to give the reader a general sense of the geographic extent of Native American tribes at the time of Euro-American contact.

(* SDS = Snoqualmi, Duwamish, and Skykomish)

geographic region (figure 8 is provided as a general guide to the distribution of tribes at the time of Euro-American settlement).

In the Sacramento Valley, along the north-central coast, and in the northern Sierra foothills of California, most Indian groups knew the California Condor by name and included it in their mythology (e.g., California Athapaskan, Yuki, Maidu, Patwin, Nisenan, Pomo, Wintun, Wappo, Coast Miwok, and Miwok Tribes) (Barrett 1908; Merriam 1910; Kroeber 1925, 1929; Loeb 1933; Bates et al. 1993). Yuki Indians (present-day Mendocino County, California) had a myth that one of the two Great Spirits "took the shape of an enormous eagle or condor" (Hodgson 2007, 295). The Nisenans and Miwoks of the central Sierra foothills believed they descended from totem animals, including the condor (Merriam 1908, 1910). The Maidus of the north-central Sierra Nevada believed that condors were spirit animals capable of bestowing shamanistic power on humans and therefore condors were never eaten or caught (Loeb 1933).

The northern California tribes embracing the Kuksu religion (Patwin, Maidu, Miwok, Valley Nisenan, Yokut, and Pomo Tribes) included condor dances as part of their annual dance cycle and tradition (Kroeber 1925). Pomo Indians, living in parts of present-day Sonoma, Lake, Mendocino, Colusa, and Glenn Counties, captured condors for their feathers and wings and may also have taken young and raised them in their villages (Loeb 1926). Some tribes in this area, including the Pomo, Maidu, Miwok, and Patwin, performed *moluku*, or condor dances, in which an entire feathered condor skin was worn (Loeb 1926, 1933; Gifford 1955). Gifford (1955) describes the process of preparing a condor skin for the dance.

When a condor is killed, the skin is kept for a dance. The hunters cut the skin from the mandible to the anus. They save the wings and the skin over the body, but not the feet or head. The body of the bird is burned, because, as the informant expressed it, the hunters feel sorry for him and do not want to see his body rot.[9] They throw tuyu seed over the body while it burns and dance around the fire and sing. The skin is stretched on sticks, and when the hunters get back to the village, it is rubbed down with deer marrow to make it soft.

---

9 According to Gifford (1955), another informant stated that only a shaman would kill a condor. Beads and seed are scattered over the body, which is buried carefully, rather than burned.

Assisted by the singer or drummer, the dancer puts on the condor skin in the chief's house, lacing it up the front of his body and sticking his legs through the skin where the bird's legs were. The condor's wings are tied to his arms and his head projects from the neck of the bird. The skin is usually so large that the tail drags on the ground. (288)

An excellent example of a condor dance costume has been preserved, collected by I. G. Voznesenskii in 1841 and housed in the Peter the Great Museum of Anthropology and Ethnography, Saint Petersburg, Russia (cat. no. 570-2; figure 9). Other tribes in the region made headdresses adorned with condor feathers (Revere 1849; Collier and Thalman 1991). As Revere noted while traveling through Lake County, California (near Clear Lake) in early fall 1846:

In the evening . . . we had a visit from a party of Indians, both male and female, attired in head-dresses composed of the black feathers of the large Californian vulture, which fell down their backs. The men were painted all over with stripes and spots, and the women wore kilts or short petticoats made of flax or hemp hackled out and fastened round the waist, but so fashioned as not to impede the motions of their limbs. They wore besides, various articles of savage finery on different parts of their persons, and all were masked. (1849, 133)

Tribes in northwest California, along the southwest coast of Oregon, and in the mountains surrounding the upper Sacramento Valley were familiar with the California Condor. Condors figured in myths, tales, and religious

Fig. 9. California Condor feather costume collected by I. G. Voznesenskii (1841). From the collection of the Peter the Great Museum of Anthropology and Ethnography (Kunstkamera), Russian Academy of Sciences, catalog number 570-2.

Fig. 10. Nora Coonskin, last shaman of the Bear River Tribe in northern California (near Cape Mendocino) holding what appear to be California Condor primaries, circa 1930 (from Nomland 1938, plate 8).

beliefs handed down by the Yurok, Wintu, Chimariko, Wiyot, Hupa, and Karuk Tribes (Curtis 1924; Harrington 1932; Du Bois 1935; Kroeber 1976; Kroeber and Gifford 1980; Sapir 2001). Condors were noted principally for potent, and sometimes dangerous, spiritual strength (Du Bois 1935). Shamans believed that confronting a condor in their dreams endowed them with spiritual power, and they used condor feathers in their attempts to cure the sick (Kroeber 1908; Curtis 1924; Sapir 2001; figure 10). In Wiyot mythology, the condor had a role similar to the biblical Adam as the ancestor of all humanity (Curtis 1924). Farther inland along the Klamath River, condors were part of Karuk mythology and language (Harrington 1932).

Condor feathers were worn by Yurok, Wiyot, Hupa, Tolowa, and Chimariko people in dance rituals, such as the White Deerskin Dance, Jump Dance, and Kick Dance (Bates et al. 1993; Paterek 1996; Sapir 2001; figure 11). Usually the condor feathers were spliced together along with flicker (*Colaptes auratus*) feathers on a rawhide strip to make especially long plumes (Drucker 1937; Bates et al. 1993; figures 12 and 13). These plumes were attached to a wooden shaft and worn atop the dancer's head (figures 11 and 13). Condor feathers were used by *pegahsoy*, or clairvoyant Karuk doctors, to get rid of shadows that caused disease (Harrington 1932,

Fig. 11. Panamenik (Karuk) White Deerskin Dance, northern California, 1910. Feather plumes in the headdresses of the deerskin dancers are made from California Condor feathers that were spliced together on a rawhide strip. Courtesy of the Phoebe A. Hearst Museum of Anthropology and the Regents of the University of California, California Indian Library Collections, Siskiyou County, book 4, number 1.

229–31; figure 12). Condor feathers were also used by doctors of the Hupa and Kato Tribes (Loeb 1932; Sapir 2001). Some Kato doctors carried a stick called a *ketaltnes*, which was stripped of bark and adorned with condor feathers (Loeb 1932). For the Hupa Kick Dance (held at the conclusion of a new doctor's training), the shaman held "a bunch" of condor feathers (Sapir 2001). Among the Yana Tribe, living between Mount Shasta and Mount Lassen, chiefs captured condors as pets (Sapir and Spier 1943).

In southwest Oregon, among the lower Rogue River Tribe, "buzzards" were considered guardian spirits with some of the most powerful curing powers (Drucker 1937). They were also believed to be patrons of hunters, who offered them the offal remaining after butchering their quarry (Drucker 1937). Farther north, Indian tribes along the lower Columbia River, the central and northern coast tribes, and Willamette River tribes were

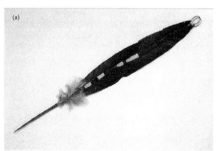

Fig. 12 (a) Hupa California Condor feather plume for White Deerskin Dance, circa 1880. Courtesy of the National Museum of the American Indian, Smithsonian Institution catalog number 001209.000; (b) Karuk California Condor feather wands, collected 1927. Courtesy of the National Museum of the American Indian, Smithsonian Institution catalog number 151922.000.

Fig. 13. California Condor feather plume (top, center) in a Hupa headdress. Courtesy of the National Museum of the American Indian, Smithsonian Institution catalogue number 4/1439. Photo by Ernest Amoroso.

likely familiar with California Condors, but loss of cultural knowledge and traditions in this area prior to ethnographic research makes understanding the true role of condors in cultural traditions difficult. Nelson Wallulatum, a Wasco chief, reported that his people kept a condor chick in camp to keep away thunder and lightning (Schlick 1994; see box 1). Condors, or condor-like figures, were also frequently depicted in basketry, handbags,

Fig. 14. Cylinder basket with condor motif, lower Columbia River, circa 1880. 9 x 5 1/4 inches diameter. Private collection. Courtesy of the Portland Art Museum, Portland, Oregon (catalog number 107). Photo by Bill Mercer.

beadwork, and stone sculptures of Columbia River tribes (Schlick 1994; Mercer 2005; Berg 2007; figure 14).

Paleontological evidence suggests that the Columbia River tribes, at least several thousand years ago, harvested condors for their feathers, and this practice appears to have taken place over thousands of years (Hansel-Kuehn 2003). However, we failed to uncover more recent evidence that condor feathers were used in the ceremonial regalia of Columbia River tribes. Although Sharp (2012) asserted that a photo of a Wyam shaman (by Edward Curtis 1910, volume 7, folio plate 20) in the vicinity of Celilo Falls showed the shaman with condor wing feathers, careful investigation of that photo suggests that he is holding a wing of a much smaller bird (P. Trail, US Fish and Wildlife Service Forensics Lab, pers. comm., 2011). Despite searching the photographic collections of Edward Curtis and the extensive photography collections at the Oregon Historical Society, as well as contacting tribal museums and several anthropologists, we did not find any photographic evidence of condor feather use among the Columbia River tribes.

We found only a few ethnographic references to condors in Washington State away from the Columbia River. Along the Washington coast, the condor appears in Coast Salish mythology as a beneficent creature living in the hills (Curtis 1913). The Chehalis people of southwest Washington feared a monster bird—which an early twentieth-century ethnographer described as "possibly a condor"—that lived near the Black River Bridge, northwest of present-day Chehalis, because they believed it was able to bite off their heads (J. Miller 1999). Condors were also mentioned among the Chehalis as having strong warrior powers (J. Miller 1999). Sharp (2012) reported several additional references from Native American tribes in Washington to birds that may have been condors. Legends of condors are notably absent from the Olympic Peninsula (Swan 1870; Pettitt 1950; Wray 2002).

In coastal British Columbia, the Sechelt Indian Band, which inhabits an area northwest of Vancouver, incorporated condors into its mythology and totem poles (Peterson 1990). Condors (*Tchass'-khain*) were seen as creatures that could use their power only for good and could counteract evil creatures but not destroy them. This contrasted with Thunderbirds (*Kwaht-kay'-ahm*), which could use their powers for destruction (Peterson 1990; see box 1). One of the stories in their oral history involved condors nesting high on a rock ledge near an impassable cliff (Peterson 1990), which is consistent with our modern understanding of condor nesting habits. The Sechelts would hunt eagles for their feathers, but not condors, which were considered protectors and guides above the hunter (Peterson 1990).

Little ethnographic evidence, and no physical evidence, of condors has been reported east of the Cascade Range apart from the archaeological dig at Five Mile Rapids, near The Dalles, Oregon. However, Klamath Indians may have been familiar with the California Condor, as they had a word (*tchuaísh*) meaning "black vulture of large size, head light-colored, or reddish" (Gatschet 1890, 446). This is contrasted with *skólos*, which clearly represented the Turkey Vulture (Gatschet 1890). Modoc and Achomawi Indians, in northeastern California, also had a name for the condor (*úm-pni*; Curtis 1924), and the Atsugewi people, also in northeastern California, had a special condor song and believed the condor brought food to the people at Ratstówni (the mythical village where the first people lived; Garth 1953). The Sahaptin-speaking tribes of north-central Oregon and south-central Washington (i.e., Yakima, Umatilla, Palouse, and Warm Springs Indians) had names for condors (*čanahúu/pachanahú*; Hunn 1990),

The Condor as the Thunderbird

Mythical Thunderbirds figure prominently in tribal stories and art throughout the Pacific Northwest, north to Alaska (Corbin 1988), leading some
to suggest a relationship between Thunderbirds and California Condors
(Smith and Easton 1964; Caras 1970; Ott 1971; D. Smith 1978; E. Clark
2003; Moen 2008). The origin of this association appears to be the popular 1964 book *California Condor: Vanishing American*, in which Smith and
Easton take creative license to weave a story describing the prehistoric,
historic, and modern relationship between condor and man.

> The condor is the thunderbird, worshipped even today by some
> ancient tribes. God-symbol, bird of life and death, he has come
> flying down through the centuries to today's man, bringing with
> him certain messages . . . a fascinating history . . . a powerful
> implication for the future. (Smith and Easton 1964, 20)

> When it thunders, the people say the condor is flapping his wings.
> The bird brings the rain. He is the thunderbird. He is good. Yet he
> brings death as well as life. (25)

This association was later picked up in the fictional story *Source of the
Thunder*, published in 1970, in which Roger Caras takes the Thunderbird
myths of several north Pacific tribes (i.e., Tlingit, Tsimshian, Kwakiutl, Co-
mox, Nootka, and Kathlamet) and ascribes them to condors. However, a
review of the myths and legends of the tribes mentioned by Caras, some
of which inhabited areas well outside the known historical range of the
condor (i.e., northern British Columbia, the Yukon, and Alaska), reveals
various Thunderbird or Thundermen stories, but no mention of condors
(Boas 1901, 1910, 1916; Swanton 1909; Sapir and Swadesh 1978).

   One of Caras's passages has been repeatedly paraphrased (see Ott
1971; D. Smith 1978; Schlick 1994; and Moen 2008):

> The Tlingit people said the condor caused the thunder by flap-
> ping its wings, even by moving a single quill. The lightning, they

Thunderbird dancer from the Kwakiutl Tribe (British Columbia coast), circa 1914. Photograph by Edward S. Curtis. The Thunderbird dancer wears a headdress representing the creature and a robe of eagle tail feathers and he imitates the bird by gesture and cry. Although some tribes associated Thunderbirds and California Condors, at least one tribe of coastal British Columbia (the Sechelt Tribe) drew several distinctions between the two. Courtesy of the Library of Congress Prints and Photographs Division, Washington, DC (reproduction number LC-USZ62-52200).

claimed, came from the bird's red eyes. An angry condor, they were sure, was likely to create thunder and lightning until it was able at last to capture a whale to carry off to a mountaintop home. (Caras 1970, 34)

However, condors are not referenced in what appears to be the inspiration for Caras's story in the authoritative text on Tlingit myths by the ethnographer John Reed Swanton (1909).

The four brothers [of a high-caste girl] now left their own village, because they said that their sister had disgraced them, and they became the Thunders. When they move their wings you hear the thunder, and, when they wink, you see the lightning. . . . At the time when these brothers first went away the people at their father's village were starving, so they flew out over the ocean, caught a whale and brought it to the town that it might be found next morning. So nowadays people claim that the Thunder is powerful and can get anything.

This Tlingit Thunderman myth is similar to other Thunderbird myths farther south (e.g., see Reagan 1917). Our review of the ethnographic literature indicates that the suggestion that North American Indian tribes in northern British Columbia, the Yukon, and Alaska had some familiarity with condors is unsupported. Occidens (1886) and Bayliss (1909) reviewed Indian myths about thunder and found that many tribes throughout North

America associated thunder with large birds, sometimes eagles or vultures, but in many cases mythical.

Some stories that have been handed down through the generations ascribe a relationship between Thunderbirds and condors in areas where we know condors historically occurred (e.g., along the Columbia River and in northern California). For example, George W. Aguilar, a Warm Springs tribal member, remembered hearing stories in his childhood of "the Thunderbird of the Columbia River Gorge [which] was the huge condor with a ten-foot wing span, the mythical ruler of storms who threw down dreadful thunderbolts, controlled all aspects of the weather, avenged wrongs, originated numerous taboos, and created volcanic activity" (2005). Nelson Wallulatum, the Wasco chief, reported that his people kept a condor chick in camp to keep away thunder and lightning (Schlick 1994). The Tolowa Tribe described the condor as "so big and powerful he can lift a whale" (Baumhoff 1958). Indeed, commonalities between Thunderbirds and condors sometimes appear striking (e.g., Thunderbirds are depicted atop totem poles with wings spread like sunning condors, and they live in caves in remote mountains, eat whales, and soar ahead of thunderstorms), suggesting that condors may have, in part, inspired Thunderbird mythology in some cultures. However, it is also possible that Thunderbirds were inspired by other, now extinct, enormous prehistoric avian scavengers (e.g., *Teratornis woodburnensis*) that may have been encountered at the end of the Pleistocene.* Moreover, many tribes familiar with condors had different names for condors and Thunderbirds, and the prominence of Thunderbird mythology well outside of the historical known range of the condor (Occidens 1886; Bayliss 1909) suggests that Thunderbirds were not direct zoomorphisms of condors in most cultures. In fact, in some areas where condors historically occurred, native inhabitants drew several important distinctions between the two, including different names and contrasting spiritual powers (e.g., Peterson 1990). Thus, while some tribes may have recognized a loose relationship between Thunderbirds and condors, the presence of Thunderbirds in tribal mythology is not a good barometer for understanding cultural familiarity with condors in the Pacific Northwest.

* Campbell and Stenger (2002) found *Teratornis woodburnensis* bones in the same stratum as human artifacts, suggesting that humans and Teratorns co-occurred in time and space; however, they did not find evidence directly linking the Teratorn remains to human occupation at the site.

as did the Okanogan Tribe in northeastern Washington and south-central British Columbia (s-ʔítwn; Mattina 1987). Farther east, in present-day north-central Idaho, northeast Oregon, and southeast Washington, condors were part of the Nez Perce lexicon (i-stá-lamkt according to Curtis 1911; qú?-nes according to Aoki 1994). East of the Rocky Mountains, the Blackfeet Indians may have occasionally observed condors, according to their oral record (Schaeffer 1951). The Crow and Gros Ventre Indians of the northwestern Great Plains also referenced condors in their mythology (Boas 1916; Lowie 1918; Ehrlich 1937; Schaeffer 1951).

Physical evidence of condor feathers in tribal regalia in the Pacific Northwest is restricted to the tribes of northern California, leaving little question that condors were historically present there, and probably locally abundant, prior to Euro-American contact. The discovery of a large number of condor bones near The Dalles, Oregon, associated with human artifacts also suggests that condors were historically present along the mid-Columbia River and were likely harvested for their feathers. Language, artwork, and oral histories suggest that condors may have been encountered, at least on occasion, in other parts of the Pacific Northwest from California to southern British Columbia, and perhaps inland to the Rocky Mountains. However, trade, along with linguistic and material cultural diffusion among Indians in the Northwest, was extensive and centered around the mid-Columbia River and its enormous salmon runs (Winther 1950; Walker 1997). Thus, without additional physical evidence, it is not possible to know whether the stories of condors handed down through oral tradition represented familiarity with condors in a tribe's home territory or knowledge gained from seasonal travels to the great salmon runs on the mid-Columbia River, where we have physical evidence of condor occurrence.

It is important to consider that in some Native American cultures, the lack of physical evidence would be expected; for example, when killing condors was taboo due to their totemic status (Peterson 1990), when feathers or other parts of a spirit animal were considered great medicines not to be seen by anyone except the possessor (Swan 1870), or when burial customs of the tribe dictated that all of a person's property should be destroyed when he or she died, as was the case with several Northwest tribes (Wickersham 1896). The lack of evidence of condors in tribal religion and rituals may also be expected in those areas where ancient traditions were lost when large numbers of Native Americans were converted to Christianity or when diseases and war ravished populations in advance of detailed ethnographic studies (Okladnikova 1983), as they did in many areas of the Pacific Northwest, especially in western Oregon and Washington (Boyd 1999).

In summary, most tribes in the Pacific Northwest had a name for the condor in their native language, and many, especially in northern California, revered them in their mythology and used their body parts for ceremonial, medical, and religious purposes. Some even believed that all living animals, including humans, were direct descendants of the condor. Given physical evidence in the form of bones and feather regalia, there is little doubt that tribes in northern California and along the mid-Columbia River had an association with California Condors. Because condors were present in their language, mythology, and art, other tribes distant from these areas may have been familiar with condors, but given the extensive trade and cultural diffusion in the region it is not possible to discern between those cultures that observed condors and those that simply knew of them from their travels to the mid-Columbia River to trade and fish for salmon.

Observation and Collection of California Condors by Naturalists, Explorers, and Settlers

Documented observations of California Condors in the Pacific Northwest by early European, American, and Russian explorers and naturalists began with William Clark, Meriwether Lewis, and other members of the Corps of Discovery in 1805 and 1806 (table 3; figure 15; appendix). During their travels along the lower Columbia River, they killed at least five condors and observed them on numerous occasions from around Celilo Falls (near The Dalles, Oregon) to the coast (Lewis et al. 2002; figure 16).[10]

In the wake of Lewis and Clark's expedition, trappers and fur traders in support of the Pacific Fur Company, American Fur Company, North West Company, and the Hudson's Bay Company were all seeking their fortunes in the region (Schafer 1909). Among these individuals, Alexander Henry, David Thompson, and Donald McKenzie reported seeing condors along the

---

10   Members of the Corps of Discovery often copied journal entries from one another. Meriwether Lewis's sketch of a condor head is shown in figure 16. William Clark copied Lewis's condor head sketch in his journal, which is housed at the Missouri Historical Society in Saint Louis (see Lewis et al. 2002; February 16, 1806, Voorhis no. 2).

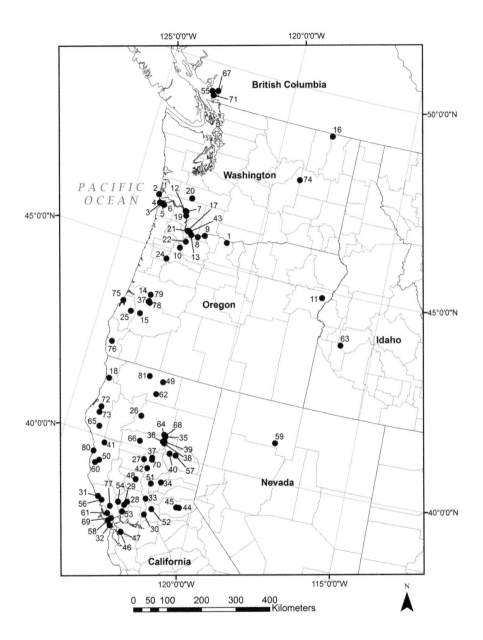

Fig. 15. California Condor occurrence records in the Pacific Northwest.
See table 3 and appendix for details.

Table 3. Records of California Condors in the Pacific Northwest, based on a review of published literature, newspaper articles, journals of early explorers and settlers, and museum records. See appendix for additional details regarding each record.

Record number	Location	Observer/ collector	Obs. type[a]	Positional accuracy[b]	Reliability score[c]	Year	Season[d]	References
1	Near confluence of the Wind River and Columbia River, Skamania County, WA	Lewis and Clark party	O	H	2	1805	Fall	Lewis et al. 2002 (weather report in Voorhis No. 4, 28 October 1805; weather report in Codex I, 29 October 1805; Clark's journal, 30 October 1805)
2	Cape Disappointment, WA	Lewis and Clark party	K[tx]	H	1	1805	Fall	Lewis et al. 2002 (Clark's journal, 18 November 1805)
3	Beach near Fort Clatsop, OR	Lewis and Clark party	O	H	2	1805	Fall	Lewis et al. 2002 (Clark's journal, 29 November 1805)
4	Near Fort Clatsop, OR	Lewis and Clark party	O	H	2	1806	Winter	Lewis et al. 2002 (Lewis's journal, 3 January 1806)
5	Youngs River, Clatsop County, OR	Lewis and Clark party	L/K	M	1	1806	Winter	Lewis et al. 2002 (Lewis's journal, 16 February 1806; Clark's journal, 16 February 1806)
6	Near Fort Clatsop, OR	Lewis and Clark party	K (x2)	M	1	1806	Spring	Lewis et al. 2002 (Gass's journal, 16 March 1806)
7	North end of Deer Island, Columbia County, OR	Lewis and Clark party	O	H	4	1806	Spring	Lewis et al. 2002 (Lewis's journal, 28 March 1806; Clark's journal, 28 March 1806)
8	Near Rooster Rock State Park, OR	Lewis and Clark party	K	H	1	1806	Spring	Lewis et al. 2002 (Clark's journal, 6 April 1806)
9	Lower end of Hamilton Island, WA	Alexander Henry and David Thompson	O	H	2	1814	Winter	Coues 1897:808
10	Above Willamette Falls, near Pudding River, OR	Alexander Henry and David Thompson	O	H	2	1814	Winter	Coues 1897:817
11	Idaho/Oregon border	Donald McKenzie	O	L	2	1818	Fall	Ross 1855:203

Record number	Location	Observer/collector	Obs. type[a]	Positional accuracy[b]	Reliability score[c]	Year	Season[d]	References
12	Near the confluence of the Cowlitz and Columbia Rivers, WA	John Scouler	K†	L	1	1825	Fall	Scouler 1905:280; National Museum of Natural History, Leiden, The Netherlands, specimen 162189
13	Near Fort Vancouver, WA	David Douglas	K†	L	1	1826	Winter	Douglas 1914:154–55; Hall 1934:5; Institute Royal des Sciences Naturelles de Belgique (Brussels) or Museum National d'Histoire Naturelle (Paris)
14	Between the Umpqua River and Willamette Valley, OR	David Douglas	O	L	2	1826	Fall	Douglas 1914:67
15	Umpqua River and south, OR	David Douglas and Norman McLeod	O	L	2	1826	Fall	Douglas 1914:216, 241
16	East of the Cascades, near the US-Canada border, WA	David Douglas	O	L	5	1826–1827	—	Douglas 1829:329
17	Near Fort Vancouver, WA	David Douglas	K†	H	1	1827	Winter	Barnston 1860:208; Douglas 1914:241; Fleming 1924:111–12; Institute Royal des Sciences Naturelles de Belgique (Brussels) or Museum National d'Histoire Naturelle (Paris)
18	Klamath River area near the present-day Humboldt-Del Norte county line, CA	Jedediah Smith	O	L	2	1828	Spring	Sullivan 1934:92
19	Cowlitz River near its confluence with the Columbia River, WA	William Fraser Tolmie	O	H	2	1833	Spring	Tolmie 1963:185
20	Cowlitz River, WA	William Fraser Tolmie	O	H	2	1833	Spring	Tolmie 1963:186

Record number	Location	Observer/collector	Obs. type[a]	Positional accuracy[b]	Reliability score[c]	Year	Season[d]	References
21	Near Fort Vancouver, WA	John Kirk Townsend	O	H	2	1834–1836	—	Audubon 1840:13
22	Willamette Falls, OR	John Kirk Townsend	K[†]	H	1	1835	Spring	J. Townsend 1848:265–67; US National Museum specimen 78005
23	Northern California [exact location uncertain—see appendix for details]	Ivan Gavrilovich Voznesenskii	K (x2)[†]; K (x2)[†x]	L	1	1840–1841	—	Blomkvist 1972:100–170; Bates 1983:36–41; Alekseev 1987; Zoological Institute, Academy of Sciences, Saint Petersburg, Russia, specimens 1583 and 1584; Museum National d'Histoire Naturelle (Paris)
24	Plains of the Willamette Valley, OR	Titian Ramsay Peale	O	L	2	1841	Fall	Peale 1848:58
25	Near Youngs River (creek) in the Umpqua Mountains, OR	Titian Ramsay Peale	O	L	2	1841	Fall	Poesch 1961:191
26	North of Redding, along the Sacramento River, CA	Titian Ramsay Peale	O	L	2	1841	Fall	Poesch 1961:194
27	Sacramento Valley between Redding and Sacramento, CA	Titian Ramsay Peale	O	L	2	1841	Fall	Poesch 1961:195
28	Valley of Napa Creek, CA	James Clyman	O	M	2	1845	Summer	Clyman 1926:137
29	Napa County, CA	James Clyman	K	L	1	1845	Fall	Clyman 1926:138
30	Between Sutter's Fort and Suisun Bay, CA	Heinrich Lienhard	O	L	2	1846	—	Wilbur 1941:42
31	Near Fort Ross, CA	William Benitz	O	M	2	1846–1847	—	E. Finley 1937:406
32	Mountains of Marin County, CA	Andrew Jackson Grayson	O	M	2	1847	Summer	Bryant 1891:52–53

Record number	Location	Observer/ collector	Obs. type[a]	Positional accuracy[b]	Reliability score[c]	Year	Season[d]	References
33	Mouth of Feather River, CA	Jacob D. B. Stillman and Mark Hopkins	K	M	1	1849	Fall	Stillman 1967:27
34	Yuba River Canyon, CA	Elisha Douglass Perkins	O	M	2	1849	Fall	T. Clark 1967:135
35	Sierra foothills, Plumas County, CA	Joseph Goldsborough Bruff party	K	L	1	1849	Fall	Reed and Gaines 1949:204
36	Mill Creek area, Tehama County, CA	Joseph Goldsborough Bruff party	O	L	2	1849	Fall	Reed and Gaines 1949:240, 245
37	Yoncalla, Douglas County, OR	Roselle Putnam	K?	L	2	1849–1852	—	Putnam 1928:262
38	Mill Creek area, Tehama County, CA	Joseph Goldsborough Bruff party	O	L	2	1850	Winter	Reed and Gaines 1949:301, 306–8
39	Mill Creek area, Tehama County, CA	Joseph Goldsborough Bruff party	O	L	2	1850	Spring	Reed and Gaines 1949:311, 325
40	West Branch, North Fork of the Feather River, CA	Samuel Seabough	K	M	1	1850–1880	—	Seabough 1880; Cummins 1893:87
41	Coast Range, Mendocino County, CA	A. K. Benton	K^{tx}	L	3	1854–1856	—	Daily Union (Sacramento) 1 April 1856
42	Near Chico, CA	Unknown	K^{tx}	L	3	1854	—	Daily Union (Sacramento) 21 June 1854
43	Near Fort Vancouver, WA	James Graham Cooper	O	M	2	1854	Winter	Cooper and Suckley 1859:141; Belding 1890:24–27
44	American River, El Dorado County (near the store of Woods & Kenyon), CA	Unknown	K	M	3	1854	Spring	Daily Union (Sacramento) 11 March 1854
45	South Fork of American River (between North and South Canyon), CA	Alonzo Winship and Jesse Millikan	L	M	1	1854	Fall	Millikan 1900:12–13

Record number	Location	Observer/collector	Obs. type[a]	Positional accuracy[b]	Reliability score[c]	Year	Season[d]	References
46	Vicinity of the redwoods of Contra Costa, CA	Joseph P. Lamson	K (x4)	M	1	1854	Winter	Lamson 1852–1861:283; Lamson 1878:152–54
47	Vicinity of the redwoods of Contra Costa, CA	Alexander S. Taylor	K[t]	M	1	1855	—	California Academy of Natural Sciences 1863, 1:71
48	Sacramento Valley, CA	John Strong Newberry	O	L	2	1855	Summer	Newberry 1857:73
49	Siskiyou Mountains, Klamath Basin, CA	John Strong Newberry	O	L	2	1855	Summer	Newberry 1857:73
50	Mendocino County, CA	Lyman Belding	O	L	2	1856–1878	—	Belding 1878, in litt.; Fisher 1920
51	Near Marysville, Yuba County, CA	Lyman Belding	O	M	2	1856–1879	Winter	Belding 1879:437; Fisher 1920
52	Near Sacramento (caught on Mrs. Harrold's ranch), CA	Unknown	L	M	1	1857	Fall	Daily Union (Sacramento) 24 September 1857
53	Lower Napa Valley, CA	Frank A. Leach	O	M	2	1857–1860	—	Leach 1929:23
54	Pope Valley, near Saint Helena, Napa Valley, CA	J. B. Wright	K	M	3	1858	Winter	Daily Alta California (San Francisco) 4 February 1858
55	Mouth of Fraser River, BC	John Keast Lord	O	H	2	1858–1866	—	Lord 1866:291
56	Russian River area of the Coast Range, CA	L. L. Davis	K[tx]	L	3	1861	—	Daily Union (Sacramento) 18 June 1861
57	Plumas County, CA	S. Stevens	K[tx]	L	3	1865	—	Daily Union (Sacramento) 25 November 1865
58	Near Marin County paper mills, CA	Unknown	K[tx]	H	3	1868	Summer	Bulletin (San Francisco) 19 August 1868
59	Winnemucca, NV	Unknown	O	H	5	1871	Summer	Daily Union (Sacramento) 26 August 1871
60	Mendocino County, CA	Unknown	K	L	3	1872	Winter	Tribune (Chicago) 24 February 1872

Record number	Location	Observer/ collector	Obs. type[a]	Positional accuracy[b]	Reliability score[c]	Year	Season[d]	References
61	Salmon Creek, Marin County, CA	Julius Poirsons	K[tx]	M	3	1873	Winter	Daily Evening Bulletin (San Francisco) 19 February 1873 [see appendix for additional citations]
62	Near the peak of Mount Shasta, CA	Benjamin P. Avery	O	H	2	1873	Fall	Avery 1874:476
63	Near the hot springs above Boise City, ID	General T. E. Wilcox	O	M	2	1879	Fall	Lyon 1918:25
64	Foothills southwest of Mount Lassen, CA	Unknown	O	L	4	1879–1884	—	C. Townsend 1887:201
65	South Eel River, Humboldt County, CA	Mr. Adams	K	M	3	1880	Spring	Bulletin (San Francisco) 5 April 1880
66	Reeds Creek Canyon, Tehama County, CA	John Bogard	K	H	3	1880	Spring	Daily Evening Bulletin (San Francisco) 7 May 1880
67	Burrard Inlet, BC	John Fannin	O	M	2	1880	Fall	Fannin 1891:22
68	Mountains south of Mount Lassen, CA	Unknown	K	L	3	1881–1882	—	C. Townsend 1887:201
69	Vicinity of San Rafael, CA	Unknown	L	L	2	1882 or prior	—	Gassaway 1882:89
70	Chico, Butte County, CA	William Proud	O	M	2	1880s	—	Belding 1890:24–27
71	Lulu Island (Fraser River delta), BC	W. London	O	H	4	1888–1889	—	Rhoads 1893:39
72	Kneeland Prairie, Humboldt County, CA	Unknown	K[t]	H	1	1889–1890	Fall	Smith 1916:205; Clarke Historical Museum, Eureka, CA
73	Yager Creek, Humboldt County, CA	F. H. Ottmer	K[t]	H	1	1892	Fall	F. Smith 1916:205; Eureka High School Hall of Ornithology, Eureka, CA

Record number	Location	Observer/collector	Obs. type[a]	Positional accuracy[b]	Reliability score[c]	Year	Season[d]	References
74	A few miles east of Coulee City, WA	Clinton Hart Merriam	O	M	2	1897	Fall	Jewett et al. 1953:160
75	Southern coast, OR	Henry Peck	K	L	3	Late 1800s	—	W. Finley 1908b:10
76	Curry County, OR	Unknown	O	L	4	Prior to 1900	—	Koford (1941), unpublished notes, Museum of Vertebrate Zoology, Berkeley, CA
77	Mountains north of San Francisco, Marin County, CA	Unknown	K†	L	1	1900–1905	—	Chicago Field Museum specimen 39613
78	Near Drain, OR	George D. Peck and Henry Peck	O	M	2	1903	Summer	W. Finley 1908b:10; Peck 1904:55; Gabrielson and Jewett 1970:180–81
79	Near Drain, OR	Henry Peck	O	M	2	1904	Spring	W. Finley 1908b:10; Gabrielson and Jewett 1970:180–81
80	Kibesillah, near Fort Bragg, Mendocino County, CA	Cecile Clarke	O	H	4	1912	Fall	Clarke, in litt. 1971
81	Siskiyou County, CA	Henry H. Frazier	O	L	5	1925	Spring	Henry Frazier, Morro Bay, CA, pers. comm. with S. Wilbur, 1971.

a   K = killed, O = observation only, L = live-captured

b   H = High (< 10 km), M = Moderate (10–50 km), L = Low (> 50 km)

c   1 = physical evidence (museum specimen) or firsthand identification based on a bird-in-hand ; 2 = firsthand observation with no physical evidence and no bird-in-hand, but with sufficient details to rule out other raptors; 3 = secondhand identification based on a bird-in-hand; 4 = secondhand observation with no bird-in-hand, but with sufficient details to rule out other raptors; proximal in time (within 10 years) and space (within approximately 100 km of physical evidence or reliable firsthand accounts); 5 = firsthand or secondhand observation with no physical evidence and no bird-in-hand, but with sufficient details to rule out other raptors; not proximal in time (within 10 years) or space (within 100 km of physical evidence or reliable firsthand accounts)

d   Spring = March–May; Summer = June–August; Fall = September–November; Winter = December–February

†   Specimen preserved (all or part) and currently in collections (see appendix for details)

†x Specimen preserved but subsequently lost or destroyed (see appendix for details)

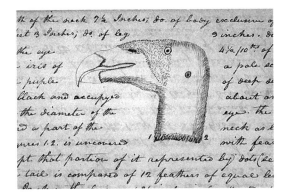

Fig. 16. California Condor drawing from Meriwether Lewis's journal (Codex J, p. 80), 17 February 1806. Courtesy of the American Philosophical Society.

Columbia River and in the Willamette Valley in the early 1800s (table 3; figure 15; appendix). As fur trapping became increasingly competitive along the Columbia River and its tributaries, new trapping grounds were sought by enterprising mountain men like Jedediah Smith. Smith pioneered a route over the Sierra Nevada and up the California coast to reach the Oregon Country (Sullivan 1934; Davis 1989). On his journey through northern California in 1828, he observed "large & small buzards," which he differentiated from eagles, ravens, and hawks (Sullivan 1934, 92). Around the same time, the Hudson's Bay Company was starting to send trapping brigades south from Fort Vancouver, along what became known as the Siskiyou Trail, into northern California as far south as San Francisco Bay.[11]

With the opening of the West to trade and travel, a few adventure-seeking naturalists, including David Douglas, John Scouler, John Kirk Townsend, and Titian Ramsay Peale, ventured into the region in the early 1800s (J. Townsend 1839; Scouler 1905; Young 1904; Douglas 1914; Stone 1916; Poesch 1961). They collected specimens for natural history museums and herbaria and recorded their observations in journals (figure 17). West of the Cascades and Sierra Nevada, these naturalists observed condors with some regularity in the first half of the nineteenth century. David Douglas described them as "common on the shores of the Columbia" and "plentiful" in the Umpqua River region in southern Oregon, observing nine

11 The trespasses and atrocities committed by the Hudson's Bay Company and other early pioneers on the Native Americans of the Rogue River Valley, along with the spread of deadly diseases, resulted in considerable friction between the natives and newcomers. Increasing use of the Siskiyou Trail by fur traders and settlers resulted in escalating violence and ultimately the Rogue River Wars, described in detail by Schwartz (1997).

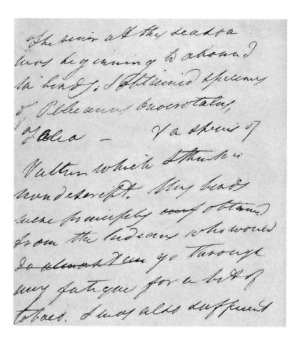

Fig. 17. Excerpt from Dr. John Scouler's journal, 22 September 1825. "I obtained specimens of *Pelecanus onocrotalus, Falco*—& a species of *Vultur*, which I think is nondescript. My birds were principly obtained from the Indians, who would go through any fatigue for a bit of tobaco." The specimen referred to is now housed at The Netherlands National Natural History Museum in Leiden. Courtesy of the Oregon Historical Society.

in a single flock (Douglas 1914). In his letters to John James Audubon, Townsend mentioned seeing condors in Oregon on a number of occasions (Audubon 1840). Titian Ramsay Peale described condors as uncommon in Oregon in 1841, but "much more numerous" in northern California at the time (Peale 1848).

As reports of the Oregon Country reached the eastern United States, churches sent missionaries to convert the Indians to Christianity and Euro-American settlers began to arrive overland via the Oregon Trail starting in the late 1830s (Doherty 2000). California was a less popular destination for settlers at that time, but by 1840 a new inland center began to emerge in the Sacramento Valley, near the nexus of the California and Siskiyou Trails, where John Sutter received a large land grant from the governor of Monterey and built a thriving trade center, ranch, and fort (Meinig 1998). Sutter provided accommodations to many travelers, at least one of whom observed condors somewhere between Sutter's Fort and Suisun Bay (M. Wilbur 1941); however, most of northern California remained only lightly populated and remote up until the California Gold Rush of 1849 (Caughey 1948).

Although most of the early fur trappers, settlers, and explorers were European or American, Russia had also been vying for a share of the Pacific Northwest fur trading industry since the early 1800s. It had established an outpost in northern California—Fort Ross—near Bodega Bay in 1812 as an agricultural base to supply its operations farther north (Alekseev 1987). By the 1830s, the Russian-American Company was winding down its operations at Fort Ross, but between 1840 and 1841, Russian scientist I. G. Voznesenskii visited the fort and surrounding areas as part of an expedition to make scientific collections for the Russian Academy of Sciences (Lipchitz 1955; Alekseev 1987). Voznesenskii's California collections are widely recognized as important to science because he was one of the last interested witnesses to provide a detailed account of California Indian life prior to the California Gold Rush and the last collector of unique objects of their material and artistic culture (Okladnikova 1983). Voznesenskii collected four California Condor specimens for the academy around San Francisco Bay and the Sacramento Valley, all of which were shipped back to Russia (see appendix; Bates 1983). With the assistance of John Sutter, he also procured a cape made from condor and eagle feathers and an entire condor feather costume from the Native Americans around Sacramento, perhaps the only one of its kind still in existence (figure 9; Bates 1983).

American immigrants continued to arrive in northern California overland via the California and Siskiyou Trails through the 1840s, but the population remained small. With the January 1848 discovery of gold in Sutter's millrace along the South Fork of the American River in the Sierra foothills, there was a surge of immigrants that would forever transform California (Caughey 1948). The sudden influx of immigrants created enormous demand for provisions and those who could provide them (Caughey 1948). With large numbers of miners and settlers infiltrating interior northern California, the entire gold-bearing region was thoroughly explored and the increasing number of local newspapers began to run stories of encounters with enormous birds with wingspans of "9 feet [and larger] from tip-to-tip," several of which were shot out of curiosity (see appendix).

By the time of the California Gold Rush, the United States and England had agreed to separate their interests in the Oregon Territory by dividing it at the forty-ninth parallel, and the Mexican government had ceded control of California to the United States. The confluence of these events, and the need for the United States to protect its newly acquired territory, created heightened interest in providing a safer, more efficient means of transportation and trade to connect the eastern and western seaboards (Albright 1921). Meinig (1998, 5) noted that:

by 1852 relatively few questioned either the necessity or the practicality of [a transcontinental railroad]. A railroad to the Pacific was the main theme of commercial conventions, orations, pamphlets, newspapers, and national periodicals, and Pacific railroad bills became the staple of every session of Congress. From this time on the basic questions were not *could* and *should* we build a transcontinental band of iron but *how* and *where* to do so.

From 1853 to 1854, Senator William M. Gwin of California got an appropriation into the War Department budget to allow the secretary of war (Jefferson Davis) to send out surveying parties along four potential transcontinental railroad routes to determine which would be best (Albright 1921). North–south routes from California to Washington were also surveyed (Baird et al. 1858; Cooper and Suckley 1859). Upon completion of the surveys, twelve illustrated volumes were published between 1854 and 1859 (Albright 1921), including detailed accounts and illustrations of the birds observed and collected along the various survey routes compiled at the Smithsonian Institution (Baird et al. 1858).

At the time of the railroad surveys, condors were apparently still regularly observed in northern California but were a rare occurrence north of California (Cooper and Suckley 1859; Jobanek and Marshall 1992; Gabrielson and Jewett 1970). Newberry (1857, 73) reported, quite poetically, seeing condors every day in the summer of 1855 while traveling through the Sacramento Valley on his way to Oregon:

> A portion of every day's experience in our march through the Sacramento valley was a pleasure in watching the graceful evolutions of this splendid bird [the California Condor].[12] Its colors are pleasing; the head orange, body black, with wings brown and white and black, while its flight is easy and effortless, almost beyond that of any other bird. As I sometimes recall the characteristic scenery of

---

12 Newberry actually used the common name "Californian Vulture" to describe the California Condor. Although the species had been described as a representative of the condor in the Northern Hemisphere (the term *condor* being originally reserved for the Andean Condor) as early as 1829 (Douglas 1829), the common names of Californian Vulture or California Vulture persisted through most of the nineteenth century among those who knew the bird. The name California Condor didn't take over in popularity until the late 1800s.

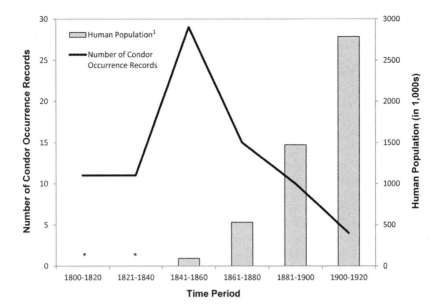

¹ Based on U.S. Census data for Washington, Oregon, and the counties of northern California (north of San Francisco). Census data represent counts at the mid-points of time periods. Due to limitations in census data, human population numbers for 1850 did not include data for the following counties in California: Alpine, Amador, Contra Costa, Del Norte, Glenn, Humboldt, Lake, Lassen, Modoc, Mono, Nevada, Placer, Plumas, San Francisco, Sierra, Siskiyou, Tehama. Due to limitations in census data, human population numbers for 1870 did not include data for Glenn or Modoc counties in California.

* Census data are not available for 1800-1840; however the human population was sparse during this time period.

Fig. 18. California Condor occurrence records and human population growth in the Pacific Northwest, 1800–1920.

California, those interminable stretches of waving grain, with, here and there, between the rounded hills, orchard-like clumps of oak, a scene so solitary and yet so home-like, over these oat-covered plains and slopes, golden yellow in the sunshine, always floats the shadow of the vulture.

However, Newberry's party reported seeing only a few condors in the Klamath Basin of California and none in Oregon (Newberry 1857). Matteson (1886) stated they were sometimes seen in southern Oregon and up along the coast. Despite the substantial increase in the number of potential observers and outlets for reporting observations in the late 1800s, the number of condor observations continued to decline across the Pacific

Northwest, with most being brief newspaper reports of condors killed or captured in northern California, or recollections of condor observations in ornithological compendiums (figure 18; appendix). James Graham Cooper—appointed surgeon on the western Pacific Railroad Surveys through Idaho and Washington, and after which the Cooper Ornithological Club was named (Emerson 1899)—first observed condors in northern California in 1855. He also noted that he had seen fewer every year since then and that unless protected they appeared "doomed to rapid extinction" (Cooper 1890).[13] After 1900, there were only a handful of condor observations in the Pacific Northwest, and the birds were completely gone from the region sometime in the early twentieth century.

Summary of Historical Distribution

The occurrence records indicate that condor observations in the Pacific Northwest appear to have been concentrated in a few areas (i.e., the Sacramento Valley, central and northern California coast, Oregon coast, Sierra Nevada foothills, Columbia River, and Umpqua foothills; figure 15). Although the distribution of these records is undoubtedly biased by the spatially concentrated distribution of observers at that time (S. Wilbur 1973; Jobanek and Marshall 1992), more than half the records were within 75 km of the Pacific coast and three-quarters were within 150 km (figure 19). This may be a result of higher food availability along the coast than in the drier intermountain basins, more favorable climatic conditions for nesting, roosting, and foraging in coastal areas, or simply an artifact of the concentration of people in areas closer to the coast. Whatever the cause, the dearth of observations farther east is probably not solely a reflection of observer bias, because the Corps of Discovery, John Kirk Townsend, and others either noted their absence east of Celilo Falls or failed to record them. Furthermore, there were several east–west travel corridors—including the Oregon and California Trails—that were used by numerous trappers, settlers, gold seekers, and naturalists throughout the 1800s (Gerlach 1970). The extensive Pacific Railroad Surveys also traversed the continent from east to west along several routes (Albright 1921). Despite the number

---

13   Cooper (1890) gave three reasons for the condor's decline: secondary poisoning, conversion of grazing land to cropland, and wanton shooting. In chapter 4 we discuss these and other possible reasons for the condor's decline.

of potential observers in the area, there were very few reliable condor observations east of the Cascade Range and Sierra Nevada, and no specimens collected from this area (table 3; figure 20). In addition to the condor occurrence data reported in table 3, there are a small number of other purported condor observations east of the Cascade Range and Sierra Nevada, but these lack credibility or sufficient details to determine their reliability:

1. Several Blackfeet Indians reported seeing large birds in current-day Montana during the 1800s, some of which could have been condors, but others of which clearly were not (Schaeffer 1951).

2. Lewis and Clark may have seen condors in Idaho, but the record is unclear. While traveling back East near present-day Weippe, Idaho, in June 1806, Meriwether Lewis indicated that "Labuish and Cruzatte returned and reported that the buzzards has eaten up a deer which they had killed butchered and hung up this morning" (Lewis et al. 2002; Lewis's journal, Friday, 13 June 1806). Other authors have referred to this observation as a Turkey Vulture (Jollie 1953; Lewis et al. 2002), although it is impossible to be sure given that Lewis and Clark used the term "buzzard" for both condors and Turkey Vultures.

3. A published report of two condors near Calgary, Alberta, in September 1896 (Fannin 1897) appears to be a misidentification of immature Golden Eagles (Brooks, in litt. 1931).

4. Moen (2008) reported oral history accounts of condors along the Yakima River (Washington), near Celilo Falls, at the eastern rim of Hell's Canyon (Idaho), and in the Clearwater River Valley (Idaho). Sharp (2012) also reported oral history accounts in the Hell's Canyon–Seven Devils area along the Snake River. However, these records are based on stories passed down as oral histories and are secondhand accounts from observers with unknown familiarity with condors; without additional evidence, their reliability is questionable (especially the later records that fall well outside of the time period of the last specimen collected in the Pacific Northwest).

The northernmost condor occurrence records are from southern British Columbia, near the mouth of the Fraser River (Lord 1866; Fannin 1891; Rhoads 1893). Chamberlain (1887) reported that condors were taken at the mouth of the Fraser River but provided no details on the disposition

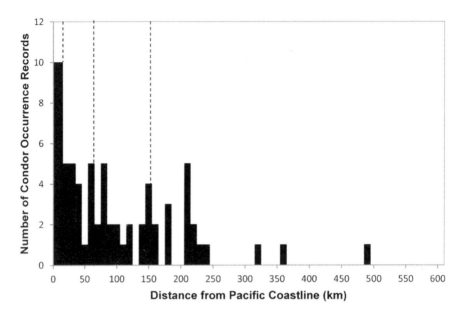

Fig. 19. California Condor occurrence records and distance (km) from the Pacific coastline. Average movement distance of breeding California Condors in southern California from 1982 to 1987 was 50–70 km (Meretsky and Snyder 1992). Vertical dashed lines represent quartiles.

of the specimens. Bendire (1892) stated that early ornithologists observed condors on Vancouver Island but did not say where or when. Although there are no definitive reports of condor specimens taken in Canada during early settlement and exploration, a single broken condor tarsometatarsus (radiocarbon dated to 2,915±15 YBP) was recently unearthed during an archaeological excavation on South Pender Island (in the Strait of Georgia, near the southern tip of Vancouver Island; I. R. Wilson Consultants, Ltd. 2006; figures 4 and 5), lending additional credibility to the Fraser River observations and Bendire's comment that condors were observed on Vancouver Island. The southern British Columbia observations are also consistent with the northern extremes of other vulture ranges worldwide (Mundy et al. 1992) and are within the northern distribution of Turkey Vultures in British Columbia (Campbell et al. 1990).[14] There are three other reports that may reference condors farther north, but these lack sufficient details to determine their reliability:

_____

14 Vultures are limited to those latitudes where there is enough daylight in winter for them to find sufficient food resources.

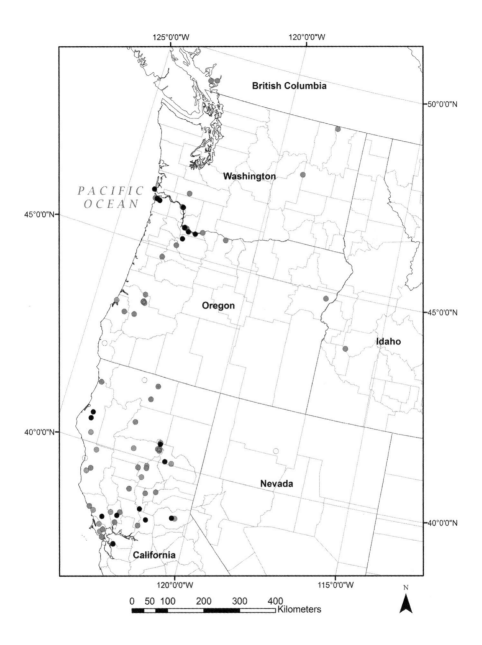

Fig. 20. Reliability of California Condor occurrence records in the Pacific Northwest. Color represents reliability class, as described in table 3 (black = 1, gray = 2 or 3, and white = 4 or 5).

1. In 1834, near Fort McLoughlin, along the central British Columbia coastline, William Fraser Tolmie observed "what I supposed a large species of vulture at the northern end [of a lake], along with some white-headed eagles [Bald Eagles, *Haliaeetus leucocephalus*] attracted probably by the dead salmon" (1963, 293). This record is questionable, although not implausible given (1) its location approximately 500 km north of other accepted observations along the Fraser River, (2) the uncertainty of Tolmie's phrasing, since he had already seen condors along the Cowlitz River in Washington the year before, and (3) the association with Bald Eagles, which may indicate that Tolmie was actually observing juvenile Bald Eagles and simply did not realize that they would have had dark heads.

2. Ross Cox, a fur trader and writer, noted shooting "a bird of the vulture tribe" northwest of the Canoe River in British Columbia in September 1817. Although this would be farther north than Turkey Vultures are typically observed (Campbell et al. 1990), lack of identifying details makes it unclear whether Cox was referring to a California Condor, Turkey Vulture, or something else.[15]

3. Sharp (2012) cited a personal communication with a Sto:lo Salish tribal biologist relating a 1935 observation of a "condor-sized bird" on the Fraser River near Spuzzum, British Columbia. In an appendix to his work, Sharp (2012) admitted that details for this record were "sparse." Given that this was approximately thirty years after the last reliable sighting in the entire Pacific Northwest, there was no definitive identification, and it was second- or third-hand information, this observation is unreliable (although not impossible).

Records of condors are noticeably absent from the Olympic Peninsula. This could reflect their actual absence or may simply be a reflection of the difficulty early settlers had accessing this remote area. The first white settlers did not arrive on the Olympic Peninsula until 1845, and in the early 1900s there were only two wagon roads in the entire western half of the peninsula (Pettitt 1950; Wood 1995). As Wood (1995, 20) put it:

---

15   Early explorers included many other birds in the "vulture tribe," including corvids (e.g., Lewis et al. 2002; journal entry of September 18, 1805).

> While the overland migrations [to the Oregon Country] were occurring during the first half of the nineteenth century, the Olympic Peninsula remained an isolated, unsought corner, a land wild, remote, and untraveled.

Historically, food resources were likely abundant for avian scavengers on the peninsula, with numerous salmon runs, pinniped rookeries, populations of elk and deer, and whale carcasses harvested by the whaling tribes in the region (Reagan 1909; Johnson and Johnson 1952; Kenyon and Scheffer 1961). Furthermore, the rugged terrain and coastal winds would likely have provided consistent slope lift for large soaring birds. However, we found no evidence of condors inhabiting the peninsula in archaeological, paleontological, ethnographic, or historical literature.

Records are also absent from the Puget Trough and San Juan Islands, despite the fact that several explorers who recorded condors along the Columbia River traveled there (Cooper and Suckley 1859; R. Miller et al. 1935; Tolmie 1963). Condors may have been rare or absent in these areas because the Puget Trough and San Juan Islands do not provide ideal soaring conditions, being relatively flat and containing large water bodies that are not conducive to thermal updrafts. The physical evidence of condors in this area is limited to the single condor tarsometatarsus found just north of the San Juan Islands, on South Pender Island (figures 4 and 5). Because this bone was found in association with cultural artifacts, it is unclear whether it was taken from a bird inhabiting the island or obtained during an excursion by Native Americans to the mainland or Vancouver Island. Cooper (1890) stated that condors ranged from "Lower California to Puget Sound" but provided no further details on specific observations within Puget Sound (Cooper 1870, 1890). In an early work, Cooper had reported that he had neither seen California Condors nor heard of their occurrence at Puget Sound (Cooper and Suckley 1859).

There are no records of California Condors in the Cascade Range. This may be the result of surveyor bias, as few early explorers ventured into the high mountains, and the Barlow wagon road over the Cascades (created to circumvent the dangerous descent of the Columbia River rapids) was not used by immigrants on the Oregon Trail until 1845 (Barlow 1902), just a few years before the last reliable observation of condors along the lower Columbia River. Those few explorers that did venture into the Cascades found rugged mountains and difficult travel, with nearly impenetrable primeval forests with dense undergrowth or canopies (Hines 1894; R. Sawyer 1932). The only report of condors in the Cascades is a secondhand account relayed in a letter from John Kirk Townsend to John James Audubon that seems dubious given the claim that condors were nesting in swamps under pine forests.[16]

> The Indians of the Columbia say that [the California Vulture] breeds on the ground, fixing its nest in swamps under the pine forests, chiefly in the Alpine country. The Wallameet Mountains [Cascade Range], seventy or eighty miles south of the Columbia [River], are said to be its favourite places of resort. I have never visited the mountains at that season, and therefore cannot speak from my own knowledge. (Audubon 1840, 13)[17]

The last record of a condor killed in the Pacific Northwest is from the mountains north of San Francisco sometime between 1900 and 1905 (table 3). Individuals familiar with the species made a few observations of condors several years after that specimen was taken, but the lack of other corroborating evidence during this time period makes these observations less reliable. One purported observation was in 1912 on the remote Lost Coast of California and another was of six condors in 1925 by Henry Frazier in Siskiyou County (table 3). A report of condors at the mouth of the Columbia River was published in the *Wellsboro Gazette* (a Pennsylvania newspaper) on 8 June 1922.

---

16 However, it is possible that some of Townsend's conversation with the locals got lost in translation. In Chinook jargon, the common language used among the lower Columbia River tribes, where Townsend was traveling, the word for "swamp" was *klimin ilêhi*, which also means "mud," or "soft ground." However, *klimin* is also the word for "broken." Is it possible that the Indians were trying to tell Townsend that condors nested in the broken trees of the pine forests in the Alpine country? The notion that condors nested in the Alpine country of the Cascade Range seventy or eighty miles south of the Columbia River is entirely plausible, as is the notion that Townsend's translation was not entirely accurate.

17 Townsend (1848) also related that condors were "reputed [almost certainly by David Douglas (1829)] to breed in the Umptqua [Umpqua] country, about fifty or sixty miles south of the Oregon Rion [?] and it is said to lay two eggs, which are entirely black!" Although it is certainly possible that condors were breeding in the vicinity of the Umpqua River, they lay only one egg at a time and the notion that their eggs were black is clearly erroneous.

The belief that the west coast or California condor, North America's largest bird is practically extinct, must be revised, for several recent news stories from towns at the mouth of the Columbia river report two pairs of the big birds are frequently seen on the rocky bluffs there. They are evidently preparing to nest later on.

The condors noticed soaring above the extensive stretch of bluffs and sandbars are very large with a wing spread of eight or nine feet. They are as black as the traditional German eagle.

However, searches of local newspapers from January to June 1922 in the archives at the Columbia Pacific Heritage Museum and in other newspaper databases failed to uncover the source of that report.

Anecdotal data can be notoriously unreliable, and evidentiary standards should become more stringent as observations move farther away from direct physical evidence in time and space (McKelvey et al. 2008). Moen (2008) and Sharp (2012) described several oral history accounts of condors in Oregon, Washington, and Idaho up until the mid-twentieth century, including the firsthand account of James Selam, who claimed to have seen two condors at Celilo Falls on the Columbia River while fishing in the 1920s. While sightings in the 1920s along the Columbia are not inconceivable, the fact that there were no other credible observations there after 1897 makes them suspect. We find the later accounts from the 1940s to the 1970s, described in Moen (2008) and Sharp (2012), to be unreliable in light of the extreme rarity and declining population trend of the condor at the time, the physical distance between the known population in southern California and the alleged sightings, and the fact that no ornithologist or birder in the region reported any sightings of this extremely rare, large, soaring species anywhere north of San Francisco after the early 1900s.

Chapter 3
Historical Movement Patterns

mi•grate (mīgrāt') *vi.* [Lat. *migrare, migrat-*] To change location periodically, esp. to move seasonally from one region to another.

Whether or not condors historically were migratory or resident in the Pacific Northwest has been a matter of debate (Koford 1953; S. Wilbur 1973, 1978) and has implications for planning future reintroductions and establishing connectivity between populations. In his seminal work on the California Condor, Carl Koford (1953) hypothesized that all condor records north of San Francisco Bay may have represented seasonal or sporadic migrants from the southern California population in search of seasonally abundant food resources. He offered an alternative hypothesis that condors were a resident species that formerly had a much larger range, but he found this hypothesis wanting due to the lack of known fossil evidence of condors at the time (Koford 1953). New information has, indeed, revealed fossil evidence in the region, ranging from San Francisco to British Columbia (table 2). Ethnographic evidence of condors inhabiting the Pacific Northwest was also unknown to Koford—although he suggested that evidence of the former occurrence of condors in Oregon might be obtained through anthropological investigations.

The migratory hypothesis was an extension of the observations and speculations of several early explorers. Douglas saw them in "great numbers on the woody part of the Columbia" but observed that they were less abundant there in winter (Douglas 1914). Douglas also stated, "I think they migrate to the south, as great numbers were seen by myself on the Umpqua River, and south of it by Mr. McLeod, whom I accompanied" (241).

Townsend reported that condors were seen on the Columbia River only during summer, "appearing about the first of June, and retiring, probably to the mountains, about the end of August" (Audubon 1840, 13). However, in a previous letter to Audubon he described the condor as "most abundant during the spring, at which season it feeds on the dead salmon that are thrown upon the shores in great numbers." This confusion is not cleared up by the occurrence records, as the only condor occurrence record from Townsend that has a definite date was the condor he shot at Willamette Falls, along the Willamette River, in April 1835.

Gambel (1847, 25) observed that condors appeared to be more numerous in California in the winter.

> [The California Vulture] is particularly abundant in California during winter; when they probably come from Oregon, as they are said to disappear from the region of the Columbia at that time.[1]

Koford (1953) acknowledged problems with the migratory hypothesis because the known records of observations in the region were asynchronous with the main salmon runs along the Columbia River, the only significant source of food that he felt could have drawn condors north.

> The only difficulty with this supposition [the migratory hypothesis] is that all of the definite records of condors in the region of the Columbia River are included between the months of September and March, the opposite of the season of the salmon run. Possibly the population of condors was so scattered in summer, when there was an abundance of food for 100 miles or more along the river, that the explorers did not notice them.[2]

---

1 Gambel's travels were limited to southern and central California (from about San Diego to Sonoma County; Beidleman 2006), so he may have observed condors becoming more abundant in California in winter, but he did not travel north of Sonoma County, California, or to Oregon, so his supposition that condors disappeared from Oregon in winter is hearsay.

2 Although Koford (1953) relied on salmon runs as the basis for his migratory hypothesis, he also expressed doubt about salmon really being a significant condor food resource. Also, he suggested that the peak of the salmon run was during the summer; actually the peak of the chinook salmon run was in September, when steelhead were also abundant. Salmon and steelhead numbers were lower from November to March (Schoning et al. 1951). However, between the various species and populations, there were salmon-steelhead runs in the Columbia almost every month of the year.

A review of the timing and distribution of known historical condor records led S. Wilbur (1973, 1978) and Sharp (2012)[3] to conclude that condors were likely a resident species in the Pacific Northwest. In the next section, we more thoroughly review migratory versus resident strategies for vultures of the world, reassess the timing of condor sightings in the Pacific Northwest with a more extensive data set of historical occurrences, and weigh the evidence for and against historical seasonal migration in California Condors. Although there are several types of migratory movements in birds (e.g., latitudinal versus altitudinal), this chapter focuses on seasonal latitudinal migration (migrating from the Pacific Northwest to southern or central California), as this has the most obvious and significant implications for reintroducing condors to the Pacific Northwest. Other forms of migration (e.g., altitudinal migration), wanderings, and dispersal movements are mentioned below, but our primary aim in this chapter is to explore whether or not at least a portion of the condors in the Pacific Northwest were resident birds.

Evolution of Migration in Vultures of the World

There are numerous factors that influence the evolution and extent of migratory behavior, including seasonality of resources, barriers, population density, historic and genetic factors, competition, mortality costs, and transportation costs (reviewed by Alerstam et al. 2003). Many of the twenty-three vulture species undertake seasonal movements, although distances traveled and the proportion of the population that migrates vary widely (Eisenmann 1963; Stewart 1977; Bernis 1983; Houston 1983; Mundy et al. 1992; Rowe and Gallion 1996; Meyburg et al. 2004; Kim et al. 2007; Mandel et al. 2008; Bildstein et al. 2009; García-Ripollés et al. 2010). Often, migration is simply an adaptation to exploiting seasonal peaks of resource abundance and avoiding seasonal resource depletion (Alerstam et al. 2003), but such movements are constrained by the length of the breeding season and barriers to soaring flight. These constraints are magnified in large-bodied vultures, which have longer breeding seasons and are unable to sustain flapping flight for more than a few minutes (Pennycuick 1968; Mundy et al. 1992).

---

3 Sharp's (2012) analysis included a number of records that are questionable given their lack of supporting evidence and distance in time or space from locations corroborated by physical evidence or firsthand accounts of trained naturalists or ornithologists.

Food availability may be the ultimate determinant in shaping movement patterns in large avian scavengers. Small vultures, which tend to be extremely flexible in their diet—an adaptation to being weaker competitors at carcasses—can exploit large and small food items and can rely primarily on small food items when necessary (e.g., Paterson 1984; Ceballos and Donázar 1990; Prior 1990). Large size in vultures is selected for only when carrion is available in large packages (Ruxton and Houston 2004), meaning that seasonal movements of large-bodied or abundant food resources, or seasonal differences in mortality rates among these food resources, are likely to be key determinants of movement patterns of large vultures (Houston 1983). However, regular seasonal movements are likely to evolve only when vultures are not physiologically constrained from being able to keep up with, or move between, seasonal food resources, and when changes in food resources are predictable (Houston 1983).

Although food availability is a strong driver of avian scavenger movements, it is not the only mechanism that can drive seasonal movement patterns. Traditional movements, possibly reflective of other physical or biological properties of the landscape (e.g., preferred roosting locations or areas with exceptional soaring potential), may also drive condor movements and may be asynchronous with food density, as S. Wilbur (1972, 1978) observed in the wild condor population of the 1960s and 1970s.

Even when predictable seasonal differences in food availability or other landscape properties exist, the evolution of long-distance seasonal migration requires that breeding individuals complete their entire breeding and rearing cycle between long-distance movements (Alerstam et al. 2003). Large birds tend to have more extended breeding and rearing periods than small birds (Blueweiss et al. 1978; Meiri and Yom-Tov 2004), making it difficult or impossible for some large birds to migrate long distances from nesting grounds on an annual basis. For those species with breeding periods longer than a year, long-distance seasonal migration is possible only for nonbreeders or juveniles. Juvenile vultures can, and often do, make long-distance movements (e.g., Arizona Condor Review Team 2002; Urios et al. 2010); apart from searching vast areas for food, juveniles may be particularly prone to wandering as they are typically outcompeted at carcasses by adults and may be seeking information about future breeding territories or location of mates. These movements may also represent attempts at dispersal, an evolutionary mechanism that may help reduce inbreeding in wild populations (Szulkin and Sheldon 2008).

The condor's breeding cycle is one of the longest among all birds (Amadon 1964; figure 21), with courtship in the late fall, egg laying from February to April, hatching from March to June, and young in nests from March to December (Koford 1953). In fact, the breeding cycle is so long that only condors that breed early in the season can breed late in successive years (Snyder and Hamber 1985). This is an extremely rare occurrence because fledgling condors remain partially dependent on their parents well into the spring after hatching. Thus, at a maximum, breeding pairs can breed two out of three years (Snyder and Hamber 1985) and chicks are dependent on parents for an extended period. Therefore, such long-distance seasonal migration among the entire population is not likely. However, nonbreeders and immature individuals are not limited in this respect, meaning that it is possible that a portion of the population could migrate (partial migration) while adult breeders remain resident.

Topography and weather, in combination with body size, can also play a role in determining whether a migratory life-history strategy is likely to evolve in soaring birds (Houston 1975). Small-bodied vultures such as Egyptian Vultures (*Neophron percnopterus*) and Turkey Vultures can travel thousands of miles seasonally and cross large expanses of water or flat terrain using flapping flight or soaring on weak thermals (Meyburg et al. 2004; Mandel et al. 2008). However, large-bodied birds, like condors, which have relatively high wing loading (a bird's weight divided by its wing area—a measure of how much lift or rising air is required to maintain soaring flight; see table 4) and weak wing musculature,[4] cannot sustain flapping flight for very long and are more reliant on areas with consistent upward air flow (e.g., mountain ridgelines) to stay aloft (H. Fisher 1946; Koford 1953; Pennycuick 1972; McGahan 1973; Houston 1975). Therefore, terrain features and weather patterns can limit seasonal movements of large soaring birds (Pennycuick 1972; Houston 1975, 1983). For example, Andean Condors are altitudinal migrants, regularly traveling large distances from mountainous nest sites to lowland feeding sites on a seasonal basis (Bildstein 2004). However, their distribution is tied to the Andes

---

4 California Condors have the weakest wing musculature of all New World vultures in relation to body weight (H. Fisher 1946).

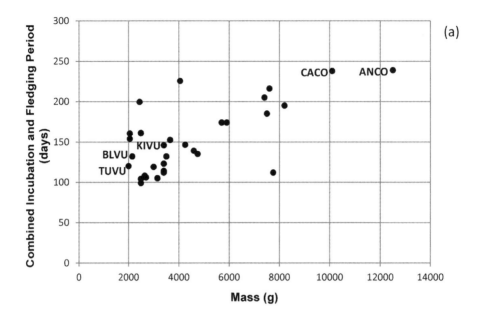

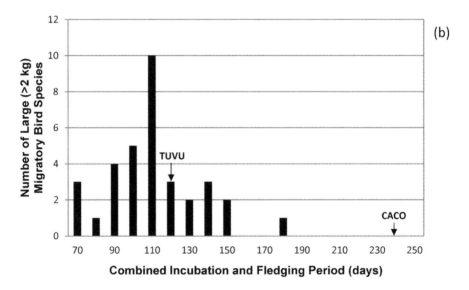

Fig. 21. (a) Combined incubation and fledging periods (days) and mass (g) for large (> 2000 g) birds of prey (both migratory and nonmigratory) with known development times (based on data in Meri and Yom-Tov 2004; Turkey Vulture data based on Kirk and Mossman 1998). (b) Frequency of combined incubation and fledging periods (days) for all large (> 2 kg) migratory bird species that have known development times (based on data in Meri and Yom-Tov 2004). Abbreviations of New World vultures (*Cathartidae*) given for reference (ANCO = Andean Condor; BLVU = Black Vulture, CACO = California Condor; KIVU = King Vulture; TUVU = Turkey Vulture). The longest development time (incubation and fledging period) reported for any large migratory bird was 171 days (Mute Swan [*Cygnus olor*]), more than two months shorter than the California Condor, at 239 days.

mountain chain and coastal areas where there is sufficient slope lift along mountain ridges or sea cliffs (Pennycuick and Scholey 1984; Lambertucci 2007). This constraint is consistent with available genetic data for Andean Condors showing detectable differences in gene distributions among populations on either side of a break in the Andes at the Northern Peruvian Low,[5] which suggests a potential dispersal barrier (Hendrickson et al. 2003). Bildstein et al. (2009) noted that Griffon Vultures (*Gyps fulvus*), another large-bodied vulture with high wing loading, had difficulties making a short (14 km) water crossing over the Strait of Gibraltar between Europe and Africa, with many attempts to cross the strait aborted. Houston (1983) also noted the dependence of Himalayan Vultures (*Gyps himalayensis*) on reliable areas of slope lift in mountainous terrain. Thus, long-distance movements of large vultures are limited to areas where slope lift or thermals provide reliable upward air currents. Mountainous regions may also be attractive to large soaring birds that have difficulty taking off over flat ground because these birds can jump off some eminence rather than struggle to reach the minimum takeoff speed necessary for flight (Pennycuick 1968).

Modern California Condor Movement Studies

Biologists have long sought to describe California Condor movement patterns in southern California (Koford 1953; A. Miller et al. 1965; USFWS 1975; S. Wilbur 1978; USFWS 1980; Johnson et al. 1983). However, these early studies were hampered by a lack of radio-tracking technology, the inaccessibility of large portions of the condors' range, and the capability of condors to make extensive movements in short periods of time. With the advent of a photographic archive of condor primary feather patterns (allowing individual identification; Snyder and Johnson 1985) and the development of lightweight patagial-mounted radio transmitters, the first quantitative study of condor movement patterns was undertaken from 1982 to 1987 (Meretsky and Snyder 1992). This study found, among other things, that (1) breeding pairs in California tended to restrict their

---

5 The Andes form a largely unbroken, high-elevation mountain chain for over 7,500 km along the Pacific coast of South America. However, there is a pronounced discontinuity in the Andes in northern Peru, where the mountain range changes direction. This region, which is about 100 km wide, is variously called the Northern Peruvian Low, Huancabamba Depression, Piura Divide, or *Depression de Huarmaca*.

Table 4. Average body mass, wing span, wing area, and wing loading of vultures.

Species	Mass (kg)	Wing span (m)	Wing area (m²)	Wing loading (Newton/m²)	Source
Lesser Yellow-headed Vulture (*Cathartes burrovianus*)	0.96	1.58	0.36	26.0	Houston (1988)
Turkey Vulture (*Cathartes aura*)	1.78	1.71	0.43	40.6	Houston (1988)
Black Vulture (*Coragyps atratus*)	1.82	1.37	0.31	57.6	Houston (1988)
Egyptian Vulture (*Neophron percnopterus*)	1.9	1.68	0.32	59	Pennycuick (1972)
Hooded Vulture (*Necrosyrtes monachus*)	2.0	1.71	0.44	45	Pennycuick (1972)
King Vulture (*Sarcoramphus papa*)	3.34	1.52	0.45	72.8	Houston (1988)
White-headed Vulture (*Trigonoceps occipitalis*)	3.97	2.16	0.63	61.8	Mendelsohn et al. (1989)
Bearded Vulture (*Gypaetus barbatus*)	5.40	2.58	0.74	71.6	Mendelsohn et al. (1989)
African White-backed Vulture (*Gyps africanus*)	5.80	2.18	0.75	75.8	Pennycuick (1972); Mendelsohn et al. (1989)
Lappet-faced Vulture (*Torgos tracheliotos*)	6.6	2.64	1.03	63	Pennycuick (1972)

Species	Mass (kg)	Wing span (m)	Wing area (m²)	Wing loading (Newton/m²)	Source
Rüppell's Griffon Vulture (Gyps rueppellii)	7.57	2.41	0.83	89.5	Pennycuick (1972)
Griffon Vulture (male) (Gyps fulvus)	7.65	2.53	0.89	84.3	Xirouchakis and Poulakakis (2008)
Griffon Vulture (female) (Gyps fulvus)	7.74	2.54	0.89	85.3	Xirouchakis and Poulakakis (2008)
Andean Condor (female) (Vultur gryphus)	8.4	2.77	0.97	85	McGahan (1973)
California Condor (Gymnogyps californianus)	8.8	2.74	0.98*	80.8	H. Fisher (1946); Ferguson-Lees and Christie (2001); Snyder and Schmitt (2002)
Cape Vulture (Gyps coprotheres)	9.29	2.57	0.85	107.2	Mendelsohn et al. (1989)
Andean Condor (male) (Vultur gryphus)	11.7	2.99	1.13	102	McGahan (1973)

*Calculated using H. Fisher's (1946) measurement of the total surface area of the bird (wings, tail, and body) and subtracting the minimum tail area as given in Ferguson-Lees and Christie (2001).

movements to within 50–70 km of their nesting site, (2) unpaired and immature condors made the longest observed daily movements, (3) foraging zones varied seasonally in accord with recent and historical patterns of food availability, and (4) most birds traveled widely among feeding zones throughout the year, probably in an attempt to maintain familiarity with unpredictable food supplies.

The analysis of spatial relationships between animals and their environment became increasingly accessible in the 1980s and 1990s with increased computing power and the commercial development of Geographic Information Systems (GIS). In conservation biology, GIS is routinely used to overlay information on the occurrence of species and the environmental variables that might explain the species' presence or absence. Using this technology and condor occurrence data from the 1980s studies, Stoms et al. (1993) evaluated condor occurrence data through time and found that the species was non-randomly associated with mapped land cover types, and the precipitous decline in the population in the twentieth century resulted in only a small reduction of the species' overall range (in southern California), as birds continued to forage over most of the range.

As captive birds were reintroduced to the wild through the 1990s and 2000s, advances in combining Global Positioning System (GPS) receivers and satellite tracking technology provided the means for more detailed studies of condor movements. Hunt et al. (2007) described the movements of condors released at Vermilion Cliffs in northern Arizona from 1996 to 2005 based on very high frequency (VHF) GPS-equipped satellite-reporting transmitters. They found that early in the release program condors moved in an unpredictable manner. During the summer months, when thermals were strong, some individuals ventured hundreds of kilometers away from the release area. For example, in July 1997 a female condor traveled 301 km to Arches National Park in Utah. In 1998, three condors traveled 387 km to Grand Mesa, Colorado, and one individual traveled 516 km to Flaming Gorge, Wyoming. Yet another long-distance flight occurred when one bird roamed as far as the Arizona-California border south of Lake Havasu City. All of these birds eventually returned to the release site. These long-distance movements were primarily by young birds (four years old or less) and apparently followed major river corridors (i.e., Colorado, Green, and San Juan Rivers; Arizona Condor Review Team 2002). As birds in the released flock became more experienced, they started to develop more predictable movement patterns, seasonally exploiting areas of higher food density (e.g., the Kolob region of Utah, where thousands of sheep are seasonally grazed in high-elevation meadows) during the warmer months when thermal updrafts were most reliable, and staying closer to the release site during the winter when thermals were less reliable and the risk of being grounded was greater.

Johnson et al. (2010) analyzed space use of reintroduced California Condors in southern California based on GPS transmitter location data from 2004 to 2009. They found that condors were beginning to recolonize the eastern portion of their historical range in the southern Sierra Nevada and were reestablishing traditional movement and foraging patterns. Unpublished data from condors being tracked with GPS transmitters in southern and central California also demonstrate that condors are reoccupying portions of their historical range, with occasional movements between the central and southern California flocks (approximately 400 km) and occasional exploratory movements from southern California into the southern Sierra Nevada (approximately 250 km; USFWS California Condor Recovery Office, unpublished data).

Seasonality of Occurrence Records in the Pacific Northwest

If condors historically expressed a seasonal migration pattern in the Pacific Northwest we would expect to see differences in the number of occurrence records across seasons. However, we found no clear north–south pattern of condor occurrence records by season (figure 22), and the occurrence records do not support Douglas's (1914) hypothesis that condors migrated south in the winter.

Condors were observed in all seasons in the Pacific Northwest (figure 23). Specific summer records in Oregon and Washington are sparse; however, John Kirk Townsend described condors variously as most abundant in the summer or spring along the Columbia. David Douglas reported them in the "summer and autumn months" as far north as the Canadian border east of the Cascades (Douglas 1914), and John Scouler obtained a condor specimen from the Native Americans along the Columbia in September 1825 (Scouler 1905). Finally, two condors were observed near Drain, Oregon, in the summer of 1903 (Peck 1904; W. Finley 1908b; Gabrielson and Jewett 1970). The rarity of summer occurrence records in Oregon and Washington is consistent with modern observational studies

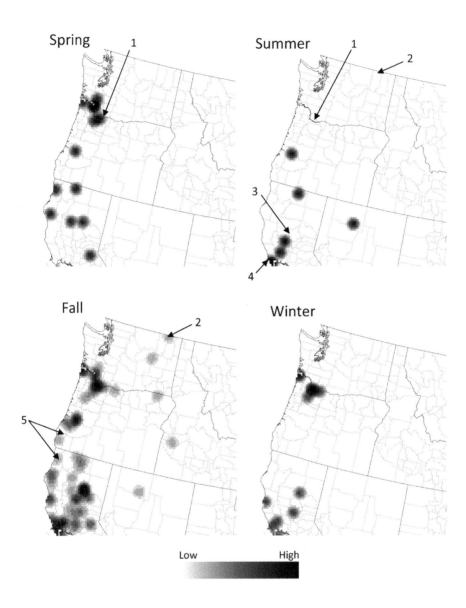

Fig. 22. Seasonal density of California Condor occurrence records in the Pacific Northwest, with annotations: (1) John Kirk Townsend variously described condors as most abundant along the Columbia River in spring or summer (Audubon 1840); (2) David Douglas reported them from the summer and fall as far north as the Canadian border (Douglas 1829); (3) John Strong Newberry reported seeing condors "every day" in the Sacramento Valley in the summer of 1855, on his journey to Oregon (Newberry 1857); (4) In the summer of 1847 Andrew Jackson Grayson observed more than a dozen condors coming to a carcass in Marin County, California (Bryant 1891); (5) In October 1826, David Douglas reported condors "in great numbers" in the Umpqua River region of Oregon, and south, observing nine in one flock (Douglas 1914).

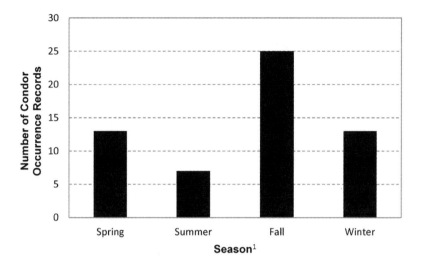

¹ Spring = March-May; Summer = June-August; Fall = September-November; Winter = December-February.

Fig. 23. California Condor occurrence records by season in the Pacific Northwest (based on data in table 3).

of Andean Condors and is possibly the result of condors migrating to more inaccessible, higher elevations in the summer in pursuit of better soaring conditions (Lambertucci 2010). Summer is also the season when some large mammal populations migrate upslope following the wave of nutrient-rich vegetation that moves up the elevation gradient through the growing season—potentially providing a more accessible and concentrated food resource for scavengers. Andean Condor movements into higher elevation sites during the summer may also be the result of the birds seeking out traditional, but only seasonally available, roosting sites, as was observed in the nonbreeding wild California Condor population in the 1960s and 1970s (S. Wilbur 1978).

The outer boundaries of condor occurrence records to the north and east are formed by fall and summer records. This is likely the result of the predominant high pressure weather patterns and high temperatures that typify the Pacific Northwest from June to September (Maunder 1968). These bring stronger thermals that facilitate long-distance movements in soaring birds. This pattern is similar to the seasonality of long-distance movements observed in condors released in northern Arizona (Hunt et al. 2007).

Fig. 24. Juvenile California Condor (US National Museum Specimen 78005) shot by John Kirk Townsend at Willamette Falls, Oregon, 1835. Gray down on the neck and head along with dark coloration on the bill and head indicate that this bird was a juvenile (≤ 3 years old), contrary to the specimen tag, which described it as an adult. Courtesy of James Dean, Division of Birds, National Museum of Natural History, Smithsonian Institution, Washington, DC.

No historical nests of California Condors have yet been confirmed in the Pacific Northwest.[6] However, two condor specimens collected in the Pacific Northwest (one held at Eureka High School, California, and the other at the US National Museum) were juveniles (figure 24). Because young condors often stay in close proximity to their nest site for the first year or two of life, it is possible that these birds were hatched in the region and that at least a portion of the condor population was resident.[7]

Summary of Historical Movement Patterns

Reconstructing historical movement patterns of condors in the Pacific Northwest is not possible given the limitations of the available data. However, we can draw the following general conclusions from the historical record, our review of migration in other vultures of the world, and data from modern condor movement studies:

---

6   Aside from the second- and third-hand accounts reported by John Kirk Townsend (Audubon 1840; J. Townsend 1848). Townsend reported to Audubon that the Indians thought they nested on the ground in the central Cascades (Audubon 1840). Townsend (1848) also reported hearsay (almost certainly from David Douglas's interactions with the Indians) that they nested in trees in the Umpqua region of southwest Oregon. Cooper and Suckley (1859) stated that "Townsend supposed he saw its nests along the Columbia."

7   While the presence of young birds in the Pacific Northwest is suggestive of breeding in the Pacific Northwest, taken alone, it is not conclusive. Condors in their third year, when they usually begin to move long distances from their natal territory, are still dark headed and often retain some down on their heads.

1. Few vultures exhibit long-distance seasonal latitudinal migrations, but many vultures exhibit shorter seasonal movements (e.g., altitudinal migration or movements associated with seasonal shifts in food supply).

2. Breeding condors cannot migrate and successfully raise young due to their extended breeding cycle, which can last for more than a year. Immature and other nonbreeding birds are not limited in this respect.

3. Condors, as large soaring obligate scavengers with high wing loading, are adapted for traveling long distances in search of food but are physiologically limited to areas with upward-moving air in the form of thermals or slope lift (Ruxton and Houston 2004). Condors attempting to cross areas of inconsistent patterns of lift risk becoming grounded and subject to mortality from predators or starvation. Thus, as long as food is available, it is generally advantageous for condors to restrict movements to areas with consistent lift.

4. Modern studies of condor movements using GPS and satellite technology have documented long-distance exploratory movements of more than 500 km, but no regular long-distance seasonal latitudinal migration has evolved in released birds. Shorter seasonal movements in relation to food availability and accessibility have been observed, but there are no discrete seasonal movements between breeding and nonbreeding areas. This is consistent with movement patterns seen in Andean Condors, a good ecological surrogate because of their size, breeding and rearing duration, and presence in a large north–south mountain chain that could facilitate migratory movements.

5. Condors were historically observed throughout the year in the Pacific Northwest and, based on anecdotal observations, do not appear to have been migratory. Exploratory movements east of the Cascades and north to British Columbia likely occurred when weather patterns were favorable.

6. Two young condor specimens were collected by naturalists in the Pacific Northwest, providing physical evidence that breeding likely occurred in the region.

Ram Papish

Chapter 4
Timing and Causes of the Condor's Range Collapse

col•lapse (kə-lăps') *vi.* [Lat. *collabi, collaps-*] 1. To fall inward suddenly: cave in. 2. To break down suddenly in strength or health and cease to function.

Population Decline

The extinction problem has little to do with the death rattle of its final actor. The curtain in the last act is but a punctuation mark—it is not interesting in itself. What biologists want to know about is the process of decline in range and numbers. (Soulé 1983, 112)

The California Condor has sometimes been portrayed as a Pleistocene relict or a senescent species that has been in a state of population decline for thousands of years (e.g., L. Miller 1942; Pitelka 1981). Despite a significant range contraction and the concurrent extinction of several large avian scavengers (including several condor species) at the end of the Pleistocene, the ethnographic, paleontological, and early observational record of California Condor populations along the West Coast of North America suggests that they were not rare at the time of Euro-American contact. While the exact population size of condors at Euro-American contact is unknown, a number of early explorers and settlers considered them common or numerous, especially in California. Bryant (1891, 52) noted that

[the California Vulture] is better known in California than elsewhere, where, previous to the civilization of that country, it was very abundant, approaching in large flocks the near vicinity of the Missions, where it contended with the coyote for the offal and carcasses of cattle slaughtered for their hides and tallow.

In the fall of 1826, David Douglas observed condors in "great numbers" along the lower Columbia River (1914). Townsend said he would not consider them numerous along the Columbia in the 1830s (Audubon 1840) but also stated, "during the spring, I constantly saw the [California] Vultures at all points where the Salmon was cast upon the shores" (J. Townsend 1848, 266). Clyman (1926) described them as being in "greate abundance" in Napa Valley, California, in 1845. John Strong Newberry (1857, 73) saw them "every day" on his travels in the Sacramento Valley in 1855. Joseph Lamson observed "upwards of fifty" condors flying over the mountains east of San Francisco in the 1850s (Monteagle 1976).

Snyder and Snyder (2005) calculated that if the 5 percent annual population decline estimated for 1950 to 1968 was extrapolated backward in time, there could have been one thousand condors alive at the beginning of the 1900s, although they recognize the possibility of differential mortality rates through time and do not consider this to be a reliable estimate of the population at the turn of the century.

While we may never know the true historical population size, the abundance of ungulates and marine mammals prior to Euro-American expansion into the region would likely have once supported several thousand condors in the western United States. A. Taylor (1859) reported that in July 1859, his friend had observed as many as three hundred condors feeding on dead sea lions in Monterey, California, but Taylor's facts were sometimes inaccurate and the identity and reliability of his source are unknown. Nevertheless, it is not beyond the realm of possibility, as flocks of several hundred vultures are not uncommon when food is abundant (e.g., Prather et al. 1976; Mundy et al. 1992; Bildstein et al. 2009).

Andean Condors—which also have high wing loading, are long lived, have slow rates of maturation, and are restricted to more mountainous regions of South America—were estimated to number roughly 6,200 individuals in 2000 (Díaz et al. 2000). Using mark-recapture techniques, Lambertucci (2010) estimated that there were 260–332 Andean Condors in just one area of northwestern Patagonia measuring 6,300 km^2 (smaller than the Olympic Peninsula). This estimate may represent a reduced number of birds from the historical population size due to ongoing threats such as persecution, poisoning, and electrocution (Lambertucci 2010). Given the similarities in body size, diet, reproductive output, and mobility between California and Andean Condors, it seems plausible that California Condors, which once occupied an area from Baja California to British Columbia, may also have historically numbered several thousand individuals rangewide. However, without additional information, this remains a tentative, untested hypothesis. Analysis of genetic data in museum specimens may provide one avenue of inquiry to evaluate this hypothesis.[1]

A number of hypotheses have been proposed for why condors disappeared from the Pacific Northwest. However, there has been no systematic and thorough accounting of the likelihood of each hypothesis for the region given the facts regarding the timing and magnitude of the threats, population decline, and range contraction (but see S. Wilbur 1978, 2004, and Snyder and Snyder 2005 for a more general discussion of reasons for the condor's historical decline). Here we assess the plausibility of each hypothesis with respect to timing, magnitude, and extent, in the hope of narrowing the range of proposed explanations.

Secondary Poisoning

The use of poison to kill predators and other mammalian pests was largely an outgrowth of the expanding fur trade and livestock industries throughout the West. Livestock expansion started during the late 1700s in the Southwest in order to provide a reliable food supply for early explorers and missionaries (Love 1916), and it reached the Pacific Northwest on the heels of the fur trading industry, which rapidly expanded in the Pacific Northwest in 1821 when the Hudson's Bay Company gained access to trapping grounds in Oregon, Washington, and British Columbia (Hammond 2006). The Hudson's Bay Company viewed competition from predators as counter to the company's interests and actively promoted wolf poisoning (Hammond 2006). It also sought to protect the small but growing number of livestock it and its workers owned. In 1839, Dr. John McLoughlin, chief factor in charge of the Columbia District for the Hudson's Bay Company, requested poison to use on the company farms in the Pacific Northwest and to sell to settlers (Hammond 2006). The company responded by sending

> a small quantity of Strychnine made up in dozes for the destruction
> of Wolves; it should be inserted in pieces of raw meat placed in such
> situations that the shepherd's dogs may not have access to them,

---

1 This work is currently underway by the authors and colleagues.

and the natives should be encouraged by high prices for the skins to destroy wolves at all seasons. (Rich 1943, 164)

Strychnine was offered for sale by the Hudson's Bay Company (Rich 1943) and was likely used around company farms at Nisqually, Cowlitz, and Fort Vancouver in Washington.[2] Titian Ramsey Peale reported that it was used to destroy wolves near Puget Sound in 1841 (Peale 1848). Missionaries also used poison to kill wolves around missions in the region as early as 1839 (Marshall 1911; Gibson 1985). In the fall of 1839, following a series of wolf and Indian dog attacks on livestock and horses near present-day Lewiston, Idaho, missionary Henry Spalding requested that the American Board of Commissioners for Foreign Missionaries send enough poison to kill one thousand wolves (Gibson 1985). The following spring he requested enough poison to kill twenty thousand wolves (Gibson 1985).

As the number of settlers to the region increased, so did interactions between livestock and predators. Following widespread wolf depredations on valuable imported livestock, predator bounties were enacted by the first provisional government in the Oregon Territory in 1843 (Brown 1892; Hampton 1997). Thomas Cox, who operated a trading post in Salem, Oregon, imported large amounts of strychnine to the Willamette Valley starting in 1847 (Minto 1905). John Minto, an early pioneer (and later a prominent sheep farmer and Oregon state representative) who arrived in the Willamette Valley in 1844, wrote:

The chief enemies of early home building were the carnivori, of which the large wolf was the most destructive, attacking all kinds of stock, colts being their most easy prey, next calves and young cattle. They kept range cattle wild and made swine band together in self defense. They ate up the first two swine I owned, and all their young but one. They ran in families most of the year, I think. I never saw more than seven or eight together, and were so voracious that they were easily poisoned, leaving the small wolf, or coyote, the most cunning and active pest. (1908, 149)

---

2 Strychnine was more formally known as *nux vomica* on some Hudson's Bay Company inventories.

Strychnine use expanded throughout the American West during the Gold Rush of 1849, when jars of it in crystalline-sulphate form quickly became a familiar feature on the shelves of trading posts (Coates 1999; Jones 2002). Roselle Putnam, daughter of Oregon pioneer Jesse Applegate, offers the following firsthand account from a settlement in southwest Oregon in 1852:

> The wolves of this country are very large and numerous[,] there has been a great many of them killed this winter, in this neighborhood with strycknine, Charles put out upwards of thirty doses of it, and I suppose every one killed a wolf at least the physician from whom we got it said it would—we have seen two die near the house—notwithstanding the quantity of poison they have taken—they are still to be heard every night or two howling round us & one impudent fellow has been in the habit of coming every night to pick up the scraps about the house & even in the porch a couple of nights ago—we gave him a dose of poison and he has not been back since. (1928, 256)

Over the course of the latter half of the century, poison would decimate wolf populations across North America, with an untold impact on scavengers (Hampton 1997). It was also used by farmers to kill smaller agricultural pests such as ground squirrels and rabbits (*Willamette Farmer* 1872).

The prevailing opinion among those commenting on the condor's status was that poisoning was to blame for the condor population decline in the late 1800s (A. Taylor 1859; H. Henshaw 1876; Cooper 1890; Lucas 1891; Streator 1888; Bendire 1892; Shufeldt 1900; *New York Times* 1906; W. Finley 1908b; Finley and Finley 1915; Lyon 1918; Tyler 1918; W. Fry 1928), yet some believed that condors and other vultures possessed some natural immunity to poisons or could disgorge the poison before it was absorbed (Scott 1936b; Harris 1941; D. Smith 1978). Opponents of the poisoning hypothesis noted the scarcity of historical accounts of California Condors actually being killed by poison (see box 2 for an account of the known California Condor poisoning events prior to releasing captive-reared birds in the 1990s).

If poisoning had been a significant factor in the condor's population decline and range contraction, one might expect more reports of condor deaths from poisoning in the historical record, as a dead bird with a nine-foot wingspan was newsworthy in the 1800s (Scott 1936b; S. Wilbur 1978, 2004). However, the absence of evidence is not evidence of absence. It is entirely plausible that poisoned birds were significantly underreported because they died in inaccessible areas, decomposed before being found, or were scavenged themselves. Moreover, those putting out the poison may have been uninterested in its side effects on what they viewed as economically unimportant birds, or they may have been misinformed and thought that condors were detrimental to livestock interests—a common misconception at the time. One trapper described the animals impacted by predator poisoning:

> Magpies, buzzards, porcupines, wildcats, skunks; Did not count them as was glad to see them dead. (Linsdale 1932, 126)

Adverse impacts on other avian scavengers with higher reproductive rates and less parental care than condors were noted. One wolf hunter in Texas reported, "Many, many hundreds of ravens were killed by eating the carcasses of poisoned wolves." Another wolf hunter in Kansas claimed that ravens disappeared entirely from the central plains from eating poisoned baits (Hampton 1997, 110).

If California Condors congregated at a poisoned carcass, something that would not be unexpected (John Pemberton documented thirty condors at a single [untainted] carcass [Koford 1953]), a large number of condors could have been wiped out in a very short time. Assuming an average of ten condors at a medium-sized carcass and a hypothetical condor population in the Pacific Northwest of one thousand individuals, just one hundred poisoned carcasses could have contaminated the entire population. The number of poisoned carcasses was actually much higher, based on accounts of settlers and travelers indiscriminately lacing carcasses with poison along trails and around homesteads. Over just a four-year period in one small area of Humboldt County, California, government trappers, farmers, and ranchers reported the loss of fifty "buzzards" (presumably Turkey Vultures) from secondary poisoning (Linsdale 1932). Linsdale (1932) also received thirteen other reports of Turkey Vultures or "buzzards" being killed by poison in California.

In parts of the world where predator poisoning is still a common practice, vulture deaths and population declines from secondary poisoning

Known Historical Poisonings of California Condors*

In 1856, the San Francisco *Bulletin* reported:

> A short time ago quite a number of the condors were found dead by the vaqueros on the Sur ranch in this county [Monterey], from the effects of eating the meat of a bear poisoned with strychnine. (*Bulletin* 1856)

In 1864, the San Francisco *Bulletin* reported:

> Some person, through accident or design, some few days since poisoned a pet vulture belonging to Dr. Canfield. The poison used was strychnine, and was probably administered in a piece of meat. The vulture was about eight months old, and measured across its wings, from tip to tip, 8 feet 9 ½ inches. (*Bulletin* 1864)

(3) Reid (1869, 377), conveyed the report of an old *ganadero* (rancher) who used strychnine to kill predators:

> There are times when certain beasts of prey, more especially wolves and coyotes become pests to the ganaderias or grazing farms; and means have to be adopted for thinning their numbers. An old ganadero, whose testimony I can trust, tells me of having employed strychnine to poison them. He did so by chopping up the flesh of several bullocks, and inpissating it with the poison. It was scattered here and there over his pastures, at places known to be frequented by the "vermin." On going one day to inspect the envenomed lure, he found not only a number of coyotes lying lifeless upon the ground, but half a dozen large vultures, that had gorged themselves on the "spiced beef." The birds were not quite dead, but only stupefied, and looking, as he said, like

pigeons that had been made drunk on wheat steeped in whiskey.
. . . The strychnine had already done most of its work; and after
fluttering awhile over the ground, now getting up, now tumbling
down again, and staggering about like so many drunken men,
one after another at length lay prostrate upon the sward, turning
stiff, almost as soon as they had ceased kicking!

Although Reid's detailed statement is contained in a fact-filled account of New World vultures, he was a novelist of children's adventure
stories and often interwove fact and fiction (Holyoake 1891). He attrib-
uted some other "facts" to his *ganadero* that are pure nonsense. Thus,
the veracity of this account is questionable.

(4) In 1880, the San Francisco *Bulletin* reported:

On South Eel river, Humbold[t] county, Mr. Adams recently poisoned a bird of the vulture species which measured nine feet
across the wings, four feet from beak to tail and eight inches from
crown to tip of beak. (*Bulletin* 1880)

(5) W. Fry (1928) documented two condors that had been killed by a
shepherd who had laced sheep carcasses with poison to kill coyotes
in Fresno County, California, in 1890 and had noted "several Condors
around the poisoned sheep the day before." Fry stated without further
support, "No doubt thousands of Condors met their deaths through
eating poison. For many years there were no restrictions placed on the
methods of poisoning [or] what kinds of poison could be used."

(6) Koford (1953) reported condors feeding on thallium sulfate–poisoned
ground squirrels around Bakersfield, California, one time observing ten
condors all feeding on poisoned carrion sometime between May and
August 1943. Condors developed a tradition of returning each year to

these same fields a few days after poisoning crews (Koford 1953). The effect of these poisoning efforts on the survival or reproduction of condors is unknown, but given reports of thallium poisoning killing Turkey Vultures and other raptors (Linsdale 1931), it is likely that it also killed some condors during its use from the 1920s to the 1940s (Koford 1953).[†]

(7) S. Wilbur (2004) summarized an incident in which three condors (two adults and one immature) were likely poisoned by strychnine at a carcass put out to kill coyotes east of Bakersfield in 1950. One adult condor was dead near the coyote carcass, and the other two birds were weak and unable to fly. After being provided with food and water for a few days, the surviving adult bird flew away. The immature bird remained in the area for eighteen days before disappearing (it was assumed to have flown away). Analysis of the dead bird revealed only a trace of strychnine in the digestive tract.

(8) Borneman (1966) reported an apparent strychnine-poisoned condor at a coyote bait in 1966. The adult condor was found in Alisos Canyon near Los Alamos, Santa Barbara County, in 1966, apparently suffering from strychnine poisoning. The bird was taken to the Griffith Park Zoo, where it was treated and subsequently released back into the wild.

(9) Snyder and Snyder (2000) reported a condor that was likely killed by sodium cyanide poisoning from an M-44 coyote trap in 1983.

---

[*] Prior to releasing captive-reared birds in the 1990s.

[†] Although Koford (1953) reports this incident of condors feeding on poisoned ground squirrels in his monograph, there is no corresponding record that he ever observed condors feeding on poisoned ground squirrels in his extensive field notes (unpublished notes, Museum of Vertebrate Zoology, University of California, Berkeley).

have been well documented. For example, predator poisoning is one of the primary causes of Egyptian Vulture declines in Spain (Carrete et al. 2007; Hernández and Margalida 2009). A total of 241 poisoning incidents were recorded in Spain, killing 456 Cinereous Vultures (*Aegypius monachus*) from 1990 to 2006 (Hernández and Margalida 2008), with as many as thirty-eight vultures poisoned in a single incident; carbofuran, aldicarb, and strychnine accounted for 88 percent of the deaths from poisoning. Reports of mass vulture poisonings within Africa are numerous (reviewed by Ogada and Keesing 2010). Mundy et al. (1992) reported 293 poisoned vultures (including White-backed Vultures [*Gyps africanus*], Cape Griffons [*Gyps coprotheres*], and Lappet-faced Vultures [*Torgos tracheliotos*]) in just a two-year period in Kruger National Park, South Africa (purposely poisoned to be sold for medical purposes on the black market). An estimated one hundred vultures were killed at a single cow carcass poisoned with strychnine near Mochudi, Botswana, in December 1984 (Mundy et al. 1992). Simon Thomsett reported 187 vultures at a single poisoned carcass in Kenya (Virani et al. 2011).

In Europe, strychnine poisoning has likely resulted in the disappearance of the Egyptian Vulture from most of the mountainous parts of the Midi in France, the loss of Bearded Vultures (*Gypaetus barbatus*) from southern Spain as well as Bosnia and Herzegovina, the decimation of several vulture species from Sardinia, the loss of Griffon Vultures from Sicily, and the extermination of Bearded Vultures and Griffon Vultures from Romania (reviewed by Bijleveld 1974).

In Africa, the use of carbofuran (which also goes by the trade name Furadan) by livestock owners to kill lions and other predators is widespread, and this powerful and toxic insecticide may be a factor in the decline of several vulture species in that region (Thiollay 2006; Ogada and Keesing 2010). This poison is particularly dangerous because it is odorless, tasteless, cheap, and readily available. In Kenya, Otieno et al. (2010) reported the deaths of twenty White-backed Vultures related to predator poisoning at a single bait site. Mineau et al. (1999) documented hundreds of cases of raptors being poisoned by the use and abuse of organophosphorus and carbamate pesticides in Canada, the United States, and the United Kingdom from 1985 to 1995. This included 125 cases of Golden Eagle mortalities from pesticide abuse related to attempts to poison coyotes or eagles. A study of secondary poisoning of eagles over a ten-year period (from 1993 to 2002) in

Saskatchewan, Canada, resulted in fifty-four putative poisoning incidents involving seventy Bald Eagles and ten Golden Eagles (Wobeser et al. 2004).

The recent catastrophic collapse of Asian vulture populations demonstrates how poisoned carcasses can quickly decimate populations of once-common avian scavengers (in this case the poison was a veterinary drug given to cattle, rather than an attempt to deliberately poison predators; Gilbert et al. 2002; Prakash et al. 2003; Green et al. 2004; Oaks et al. 2004; Shultz et al. 2004). In less than ten years, two species of *Gyps* vultures declined by over 90 percent in India (Prakash et al. 2003) due to diclofenac poisoning (Oaks et al. 2004; Shultz et al. 2004). The sociality and large range size of many vultures means that massive population declines can occur with seemingly small percentages of carcasses poisoned. For example, Green et al. (2004) reported annual mortality rates of 22 to 50 percent in *Gyps* vultures in Southeast Asia, which could be explained by diclofenac poisoning in less than 1 percent of carcasses.

Although direct evidence linking poisoning and California Condors is limited, the circumstantial evidence that condor declines may have been caused by poisoning is significant. Most notably, vulture declines in other parts of the world have been directly linked to predator poisoning campaigns or inadvertently poisoned carcasses, something that was not known at the time this hypothesis was roundly criticized by Scott (1936b), Harris (1941), Koford (1953), and S. Wilbur (1978). Socially foraging vultures are particularly susceptible to secondary poisoning because a large portion of the flock can be poisoned at a single carcass. Given the evidence now available, it seems plausible that poisoning played a significant role in the condor's range contraction. The dearth of reasonable alternative hypotheses (aside from direct persecution through shooting, and possibly lead poisoning [see below]) lends further support to this idea.

Lead Poisoning

Lead toxicosis is currently considered to be the anthropogenic threat of primary concern for reestablishing viable self-sustaining condor populations (Wiemeyer et al. 1988; Pattee et al. 1990; Meretsky et al. 2000; Snyder and Snyder 2000; D. Fry 2003; Cade et al. 2004; Cade 2007; Woods et al. 2007; Finkelstein et al. 2010; Walters et al. 2010; Finkelstein et al. 2012; Rideout et al. 2012). The most likely pathway for lead ingestion is through gut piles left in the field by hunters after they remove the meat, or through animals that are shot and unrecovered (Church et al. 2006). Gut piles and carcasses can contain significant quantities of lead fragments from spent ammunition (Hunt et al. 2009), and numerous studies have shown elevated lead levels in avian scavengers, including condors, vultures, eagles, and ravens, with the most likely source being lead ammunition (Janssen and Anderson 1986; Pattee et al. 1990; Church et al. 2006; Cade 2007; Craighead and Bedrosian 2008; Bedrosian and Craighead 2009; Helander et al. 2009; Stauber et al. 2010; Lambertucci et al. 2011).

Although lead is a well-known neurotoxin and nephrotoxin (a toxin having specific destructive effects on kidney cells), the threat of lead toxicosis to California Condors from spent ammunition became recognized only after Locke et al. (1969) discovered that Andean Condors were susceptible to lead poisoning and suggested that California Condors might also be susceptible. Additional research in the 1980s confirmed that California Condors were ingesting lead in the wild and were susceptible to its effects (Janssen and Anderson 1986; Snyder and Snyder 1989). Given the relatively recent realization that lead ammunition is a threat, the extent of its historical impact is not well understood.

Changes in bullet construction and concomitant changes in bullet velocity may have changed the bioavailability of lead to scavengers beginning in the 1890s with the invention and mass production of smokeless powder cartridges to replace the slower burning and less powerful black powder cartridges. Smokeless powder rounds and the rifles designed specifically to use these new rounds were immediately popular. The Winchester Model 1894, the first commercial rifle designed to chamber smokeless powder rounds—specifically, the .30-30 Winchester, or .30 WCF, which is still a popular round—was a sensation, with over one million sold by 1927 (T. Henshaw 1993). Numerous other arms and ammunition manufacturers were also developing smokeless powder munitions and guns to shoot them by the time of the First Annual Sportsman's Exposition at Madison Square Garden, New York, in 1895 (*Forest and Stream* 1895).

Aside from the obvious advantage of making less smoke, smokeless powder rounds also generated significantly more force than black powder (Kneubuehl et al. 2006). Maximum muzzle velocities for black powder rifles were restricted to about 1,850 feet per second (ft/sec) and typically fell within the range of 1,250–1,450 ft/sec (Walter 2006). The significant increase in energy created by smokeless powder was not effectively harnessed by conventional solid lead bullets, which led manufacturers to quickly develop a jacketed bullet, consisting of a lead core encased in a "jacket" of copper, cupronickel, tin-plated copper, or mild steel. Jacketing the lead bullets decreased the friction of the bullet as it traveled down the barrel, reduced leading of the rifle barrel, and greatly increased muzzle velocities (i.e., resulting in loads exceeding 2,500 ft/s; Whelen 1918). With greater velocities, bullets could be made longer and thinner to fly faster and stay true at greater distances. Early full-metal jacket bullets (where the lead core was fully encased in another metal) were found to be unsatisfactory for big game hunting because they tended to penetrate straight through the animal without causing sufficient damage, expending most of their energy beyond the target. To rectify this problem the cartridge companies soon developed the soft point bullet, a jacketed bullet with a small amount of exposed lead at its tip. These bullets expanded in animal tissue, causing serious wounds (Whelan 1918).

Despite their advantages, higher velocity jacketed and soft tip bullets have a much greater rate of fragmentation than solid lead bullets fired from a black powder rifle (Whelen 1918). The fragmentation of bullets into a cloud of small particles increases the surface area of metallic lead and its uptake by animals that consume the fragments (Pauli and Buskirk 2007; Hunt et al. 2009). Conversely, when bullets remain largely intact (as the lead bullets from black powder rifles likely did), they may exit the animal without significant fragmentation, thus leaving little to no lead for uptake by scavengers (Pauli and Buskirk 2007), or the intact bullet may be large enough for scavengers to detect and avoid while feeding (e.g., Stendell 1980).

To date, there has not been a comparative test of historical firearms and ammunition to see whether lead bioavailability to scavengers in contaminated carcasses may have changed since the 1800s due to changes in firearms or bullet construction.[3] Such a test may help clarify the magnitude of lead poisoning as a cause of the historical condor population decline in the Pacific Northwest. Tests of lead isotopes in feathers or bones from museum specimens may provide a direct measure of historical condor lead exposure.

If bullets shot from historical firearms did fragment and contaminate carcasses, lead poisoning could have been a significant cause of condor

---

3 This research is now being performed by Dana Sanchez, Clint Epps, and David Taylor at Oregon State University.

mortality in the late 1800s due to intensified market hunting at that time on native ungulates and pinnipeds (see "Reductions in Food Availability" section, below), as well as increased subsistence hunting from the growing human population. Shotgun pellets from the extensive market, recreational, and subsistence killing of waterfowl (especially in and around the Central Valley of California) and other small game may also have been a source of lead contamination, but condors are rarely observed feeding on bird carcasses due to their small size. Shotgun pellets have been implicated in some modern condor deaths (Rideout et al. 2012), but the source of that lead shot is unknown (some possibilities include upland game bird hunting, plinking small mammals, dispatching sick or injured livestock or pets, or depredation kills). Regardless of whether additional research clarifies the role of lead poisoning in the early decline of the condor, the overwhelming evidence that modern lead ammunition is the primary threat to reestablishing self-sustaining condor populations (Walters et al. 2010; Finkelstein et al. 2012; Rideout et al. 2012) means that reducing the threat of lead-contaminated carcasses is essential for a successful reintroduction program in the Pacific Northwest.

DDT, DDE, and Eggshell Thinning

DDT was first developed by an Austrian graduate student in 1873, but its properties as an insecticide were unknown until around 1940, when a Swiss chemist, Dr. Paul Hermann Müller, discovered that it was effective at killing insects—a discovery that later won him the Nobel Prize in medicine (see http://nobelprize.org/nobel_prizes/medicine/laureates/1948/muller-bio .html). DDT was widely used as a pesticide during World War II, with peak US production in the 1960s (Woodwell et al. 1971). Following the war, the Montrose Chemical Corporation of Los Angeles, California, the largest manufacturer of DDT in the world, developed a waste disposal system that funneled DDT and the plant's processing wastes into the county sewer system, which drained into the ocean and contaminated the Southern California Bight (Kehoe and Jacobson 2003).

DDT and its metabolite DDE are known to bioaccumulate through the food chain and cause eggshell thinning, increased incidences of breakage, and resultant declines in productivity in birds. High levels of DDE were reported from condor eggshells in the 1960s, and DDE has been suspected as a possible cause of decreased productivity in the historical population

(Kiff et al. 1979; Kiff 1989; but see Snyder and Meretsky 2003) and in condors that currently feed on migratory sea lions along the California coast (J. Burnett, Ventana Wildlife Society, pers. comm., 2009). While eggshell abnormalities in condors feeding on marine mammals are not disputed, whether or not these abnormalities are caused by DDE or some other unaccounted for contaminant in the marine environment has been questioned (Snyder and Meretsky 2003; N. Snyder, pers. comm., 2012).

DDE was not present in the environment when condors inhabited the Pacific Northwest and therefore was not associated with their historical decline and range contraction throughout the nineteenth century. However, given eggshell abnormalities in condors feeding on marine mammals, future reintroduction efforts should consider the scope and magnitude of DDE and other potential contaminants (e.g., mercury, polychlorinated biphenyls [PCBs]) in marine mammals and surrogate coastal scavengers when evaluating potential reintroduction sites in the Pacific Northwest.

Collecting and Shooting

Collecting condor specimens, as well as wanton shooting, has been suggested as a major cause of the condor population decline in the 1800s and early 1900s (Dawson 1923; Scott 1936a, 1936b; Koford 1953; McMillan 1968; S. Wilbur 1978). Reasons for shooting condors were varied and included general curiosity, target practice, perceived protection of livestock, use of quills as gold-dust containers or for making tobacco pipes, and the desire to obtain specimens for private or institutional museum collections (reviewed by S. Wilbur 1978; also see Cooper 1890; Bidwell 1890; Anthony 1893; Douglas 1914; Scott 1936a; Harris 1941; Koford 1953). David Douglas noted that in the Oregon Territory circa 1826, condor quills were "highly prized by the Canadian voyageurs for making tobacco pipe-stems" (1914, 154–55). However, the extent of killing to obtain these pipe stems is unknown.

Condor skins were collected in the Pacific Northwest for museums, beginning with Dr. John Scouler in 1825.[4] By the dawn of the twentieth century, condor skins were extremely valuable and regularly sought by

---

4 Lewis and Clark's party killed a number of California Condors but apparently preserved only a few parts, which were deposited in Peale's Museum in Philadelphia. The museum was disbanded in the late 1840s and the collections were sold off, and it is unknown if these parts are still in existence (S. Wilbur, pers. comm.).

collectors (Blake 1895; S. Wilbur 2004). Koford (1953) estimated that approximately two hundred condors and condor eggs had been taken as specimens from the late 1700s to the mid-1900s. S. Wilbur (1978) used an extensive survey of museum collections and early diaries and journals to document 177 condors killed and 71 eggs taken for collections. Additional research since 1978 has revealed an additional 18 condors killed, for a total of 195 (S. Wilbur, pers. comm., 2011).

With the rise of zoological institutions, there was also a growing demand in the late 1800s and early 1900s for live specimens. Thus, condors were captured and shipped to zoos in London, New York, Philadelphia, and Washington, DC (Sclater 1866; W. Finley 1910; Dixon 1924; see box 3). Condors were also live-captured as curiosities to display in hotel courtyards, or even to be kept as personal pets (*Daily Union* 1857; Gassaway 1882; Holmes 1897; Daggett 1898; Rising 1899; Stephens 1899; Millikan 1900).

In addition to shooting for collections, shooting without any apparent purpose was also a relatively common occurrence. S. Wilbur (1978) documented forty-one condors shot for no apparent reason between 1806 and 1976. As W. Finley (1941) noted: "because of its size, it is a mark for wanton hunters with long-range rifles who often penetrate the wilder mountain regions where this bird lives." J. G. Cooper (1890) stated:

The other [reason for the condor's population decline, in addition to poisoning and reduction in livestock abundance in southern California] is the foolish habit of men and boys, who take every opportunity of shooting these birds, merely because they are so large and make good marks for their rifles when they want to practice at vultures' heads as a preparation for the annual turkey shooting in the fall. Some may even believe that the vultures may injure their live stock, but with little reason.

Some killing may have been done as retribution for appropriating game. Andrew Jackson Grayson noted that prior to 1847, the condor was "much disliked by the hunter for its ravages upon any large game he may have killed and left exposed for only a short length of time" (Bryant 1891, 52–53). Although not specifically stated, this suggests that some hunters

An Early Captive Condor in Oregon

William Finley's juvenile pet condor, General, on his perch along the Willamette River, Oregon, summer of 1906. Photograph by Herman T. Bohlman and William L. Finley. Courtesy of the US Fish and Wildlife Service, National Digital Library.

On 10 March 1906, William L. Finley and photographer Herman T. Bohlman found a condor nest near Pasadena, California, containing an egg (W. Finley 1910). The egg hatched on 22 March 1906, and Finley and Bohlman returned eight times to document and photograph the development of the young condor, which they named General (W. Finley 1908a; Finley and Finley 1915). In early July 1906, Finley took the young bird captive and brought him back to Oregon, where the condor resided for two and a half months (W. Finley 1908a, 1910). During that time Bohlman took photos documenting General's development and captured playful moments as Finley and his wife, Irene, treated General as their

Unidentified man feeding General at Jennings Lodge on the shore of the Willamette River, Oregon, in the summer of 1906. Photo by Herman T. Bohlman and William L. Finley. Courtesy of US Fish and Wildlife Service, National Digital Library.

pet. By mid-September, General was beginning to take short flights out over the Willamette River and the Finleys began to worry that their pet condor would fly off or be shot (Matthewson 1986).

On 28 September 1906, General was shipped by rail to William T. Hornaday, the director of the New York Zoological Park (New York Zoological Society 1907; W. Finley 1910; Matthewson 1986), where he lived until 1916, when he died of unknown causes. Finley retained an affection for his pet condor, publishing a number of popular articles about him (W. Finley 1908a; Finley and Finley 1915) and visiting him at the zoo in New York on at least two occasions (W. Finley 1910).

around that time killed condors in retribution for taking their game. Many of these killings likely went unreported.

In the Pacific Northwest, we documented thirty-nine condors killed for museum collections or shot for no apparent purpose from 1804 to 1905 (table 3). In addition, two condors were live-captured and kept as pets.[5] This is a relatively small number of individuals over the course of a century, even for a long-lived vulture. However, there are surely many shooting deaths that went unreported. Shooting animals for entertainment or target practice was not considered morally objectionable at the time. As noted by John Work, who led a Hudson's Bay Company fur brigade through northern California in 1832–1833:

> When the most of the people have ammunition and see animals they must needs fire upon them let them be wanted or not. (Maloney 1945, 31)

Therefore, while we can document only a small number of condors actually killed in the Pacific Northwest, our imperfect knowledge regarding actual numbers taken and our recognition of the lack of moral impediments to killing condors leaves the possibility open that this was a significant threat to the population and a primary factor in the extirpation of the condor from the Pacific Northwest.

Egg Collecting

Egg collecting did not take hold in North America until the 1860s and reached its zenith from 1885 to the 1920s (Kiff 2005). In combination with shooting, egg collecting may have had an impact on California Condor populations in southern California (S. Wilbur 1978). However, given that condors will double clutch if the first egg is removed or destroyed (Snyder and Hamber 1985), the extent of this impact on population demographics is unclear.

Regardless of the extent of the impact of egg collecting in southern California, there are no certain records of condor nests or eggs collected

---

5 A third condor was injured and kept as a pet for some time, escaped once, was found "more dead than alive," later escaped again, and was never found. See Millikan (1900) for this fascinating account. An abbreviated version of this account is also provided in the appendix, record 45.

in the Pacific Northwest. In 2003, the Napa-Solano Audubon Society published a note in their book *Breeding Birds of Napa County*, indicating that
a condor egg was collected in Napa County on 16 August 1845. However,
the veracity of this report is questionable given the late summer collection
date. Other than this dubious record no condor eggs are known to have
been collected from the Pacific Northwest. Thus, egg collecting was not a
factor in the condor's extirpation from the region.

Loss of Nesting Habitat

California Condors, and New World vultures in general, have variable nesting habitats, using large tree cavities, caves, crevices in rock faces, and rock
ledges (Snyder et al. 1986). Although historical nesting sites of condors
in the Pacific Northwest were not documented,[6] they are known to nest
in coast redwoods (*Sequoia sempervirens*) near Big Sur, California (Ventana
Wildlife Society, unpublished data), and historically nested in giant se-
quoias (*Sequoiadendron giganteum*) in the southern Sierra Nevada (Snyder
et al. 1986). Accordingly, they may have used cavities in large trees from
northern California to southern British Columbia. Therefore, it is worth
considering whether early logging of old-growth forests may have substantially reduced the number and distribution of suitable tree nesting sites for
condors at the time condor populations were declining in the region.

The earliest logging of redwood forests was by the Spaniards near San
Francisco Bay, but these were small operations (Browne 1914). Up until
the Gold Rush era, starting in 1849, most lumber for southern California
was still shipped from New England; Oregon and Washington had only a
few small mills that supplied lumber locally and to the Hawaiian Islands
(Ficken 1987). Commercial logging of the coast redwoods started around
1850 with the increasing demand for lumber associated with the Gold
Rush, first near San Francisco and soon thereafter around Humboldt Bay.
Redwood groves around San Francisco were heavily impacted by logging

---

6   Aside from the second- and third-hand accounts reported by John Kirk Townsend
(Audubon 1840; J. Townsend 1848). Townsend reported to Audubon that the Indians
thought they nested on the ground in the central Cascades (Audubon 1840). Townsend
(1848) also reported hearsay (almost certainly from David Douglas's interactions with the
Indians) that they nested in trees in the Umpqua region of southwest Oregon. Cooper and
Suckley (1859) stated that "Townsend supposed he saw its nests along the Columbia."

following the Gold Rush, and few remained south or east of San Francisco Bay by 1880 (Schrepfer 1983).

Outside of the San Francisco Bay area, logging in the Pacific Northwest was highly localized through the mid-1800s due to limited transportation infrastructure and technology (McKelvey and Johnson 1992; Rajala 1998). Early logging operations relied on handsaws and axes to fell trees and oxen to move logs along primitive skid roads (Pomeroy 1965; figure 25). Loggers were selective, restricted to areas in close proximity to a waterway or sawmill, and largely limited to supplying local markets (Rajala 1998). Large-scale mechanized clear-cut logging operations and the expansion of the forest industry into continental and Pacific Rim markets were not in place along the Pacific Northwest coast until transcontinental railroad lines were completed (the Northern Pacific line was completed in 1883, and the Great Northern was completed in 1893; Rajala 1998; Hessburg and Agee 2003), and by this time, condor sightings in the region were exceedingly rare. Moreover, the trees that condors selected to nest in were not the trees that lumbermen preferred because condors require large cavities or broken tops in which to nest and nurture their eggs and young.

Fig. 25. Loggers near the lower Columbia River, circa 1905. Photo by J. F. Ford. Courtesy of the Oregon Historical Society, negative OrHi 37858.

We cannot rule out the possibility that some nest trees were harvested with eggs or young in them, especially in the redwoods south and southeast of San Francisco—where there is evidence that condors were nesting. However, condors are known to renest if they lose their first egg early enough in the breeding season (Snyder and Hamber 1985), are not bound to a specific nest site from year to year (although some sites are used year after year), and invest virtually no energy in nest site preparation (Snyder et al. 1986). Moreover, alternative nest sites, including remnant stands of large trees, rock ledges, and caves, are locally abundant in the region, especially in the more remote mountainous areas that are typical of condor nesting habitat. Finally, condors are long lived and could have persisted in the region for decades despite decreased productivity as long as adult mortality was not too high. Thus, even if a number of tree nests were lost to early logging in the region, we would not expect this loss to have caused a dramatic population decline or local extirpation.

Native American Ritual Killings

Many tribes in the western United States used condor feathers, bones, and parts for ceremonial or spiritual purposes (see chapter 2). While some tribes sacrificed condors to adorn dancers in entire condor cape suits (Loeb 1926; Gifford 1955), others believed that killing a condor was taboo (e.g., Peterson 1990). Despite Native American ritual use by some tribes, condors persisted in the region for millennia prior to Euro-American immigration. The question, then, is whether ritual killings were increasing around the time of the condor declines in the mid-1800s to such an extent that they could have substantively contributed to the extirpation of condors from the Pacific Northwest, as some have suggested (McMillan 1968; Snyder and Snyder 2000).

In 1825 there may have been as many as 180,000 Native Americans inhabiting the Pacific Northwest, with 30,000 below the confluence of the Columbia and Snake Rivers (Carey 1935; Winther 1950) and 70,000 along the north coast of California (Cook 1956). But between 1829 and 1832, a series of fatal epidemics had a devastating effect on the native people, especially along the lower Columbia River (90 percent population decline from 1805 to 1855) and the Sacramento Valley—such that families, whole villages, and even entire tribes were destroyed or so depleted that they lost their identity and became merged with others (*New York Times* 1874; Carey 1935; Boyd 1999).

In addition to disease, battles in northern California and southwestern Oregon between Native Americans and settlers and miners culminated in the death or forcible removal of most Native Americans from this region to reservations in 1856 (Bledsoe 1885; Mark 2006). Those tribes that remained in scattered villages or rancherias had been greatly diminished by wars with settlers and disease epidemics (Bledsoe 1885; Nomland 1938; Boyd 1999; Mark 2006). By the 1870s, the majority of survivors had been moved to reservations, sometimes well outside of their historic environment, and pooled with the remnant members of other near-extinct groups (Suttles 1990).

By 1900, only a few thousand Native Americans remained in the entire Pacific Northwest (Boyd 1999). Little information on interactions with wildlife survived the drastic reduction in Indian numbers and the disruption of traditional activities. Although we do not have data on exactly how many condors were killed for ritual purposes through time, given the backdrop of disease, wars, massive population reductions, and the forcible removal of Native Americans from their homelands, it seems highly unlikely that condors in the Northwest were experiencing increased mortality from ritual killings in the mid-1800s that would have substantively contributed to their extirpation from the region.

Even if the remnant Native American populations wanted to use condors in their ceremonies, condors were becoming increasingly rare, making them harder to find and kill. Loeb (1926) reported that the Pomo Indians in California, who traditionally killed condors and used the entire skin in their condor dance, had stopped using condor skins in the early 1900s and instead merely imitated their appearance with other materials.

Although some have asserted that Native American ritual sacrifice substantially depressed condor populations, leading to continuous overall population declines long before the arrival of Europeans (McMillan 1968; Snyder and Snyder 2000), the magnitude and extent of this impact is unclear. A review of the ethnographic and archaeological record indicates that condors were killed in annual rituals in some areas—most notably in and around the Sacramento Valley and in southern California (Harris 1941; S. Wilbur 1978; Simons 1983). However, tribes in other areas revered the condor, and killing it was taboo (S. Wilbur 1978).

Food Increases in Central and Southern California

In 1769 Spain sent Friar Junípero Sérra and a small band of Franciscan monks to settle Alta California, establish missions, convert the natives to Christianity, and claim the land in advance of the Russians, who from 1741 to 1767 had been sending expeditions from Alaska southward along the coast (McRoskey 1914). By 1804, nineteen missions had been established from San Diego to San Francisco, with two more missions constructed by 1830 (McRoskey 1914). In 1834, only sixty years after the friars came north from Baja California with two hundred head of cattle, there were at least twenty thousand Native Americans at the missions and approximately eight hundred thousand cattle, horses, sheep, and mules (McRoskey 1914). The main industry of the missions was the tallow and hide trade, which resulted in an abundance of offal for scavengers.

> The hides and tallow are the only parts exported, the dried beef being consumed in the country as well as the finer quality of tallow or "manteca," made from the fat of the intestines. The heads, horns, hoofs, bones, &c., are utterly wasted and thrown away; and, indeed, until within a few years, immense numbers of cattle were slaughtered for their hides alone, the entire carcass being left to corrupt, or feed immense numbers of wild beasts and large vultures, which were thus greatly encouraged and augmented. (Revere 1849, 100)

The Mexican government's threat to secularize the Spanish missions caused the slaughter of about one hundred thousand cattle in 1834 (Cronise 1868). Koford (1953, 68) noted that "this was the year that the last specimen of the condor was taken on the Columbia River. If condors had migrated northward in search of food, they no longer had to do so."

Koford's quote suggests that the massive slaughter of cattle in 1834 might have resulted in condors abandoning traditional foraging grounds in the Columbia River region. However, evidence now suggests this hypothesis should be rejected for the following reasons:

1. At least a portion of the population was likely resident in the Pacific Northwest (see chapter 3).

2. John Kirk Townsend killed a juvenile condor (suggestive of breeding nearby) near Willamette Falls (approximately 20 km south of the Columbia River) in the spring of 1835, a year after the slaughter.
3. A juvenile condor was killed near Eureka, California, in the fall of 1892.
4. Numerous reliable condor observations were documented throughout the Pacific Northwest in the second half of the nineteenth century (see table 3), including an observation along the mid-Columbia River in 1850 (Cooper and Suckley 1859) and an observation near Coulee City (which is on the Columbia River in Washington State) in 1897.

Reductions in Food Availability

The leading hypothesis for the prehistoric range contraction of condors is that food resources became limited in the interior of the continent and farther east at the end of the Pleistocene, causing condors to become restricted to the West Coast (see chapter 2). Although some have speculated that food reductions may have been a major cause of the species' population decline and range contraction in the nineteenth and twentieth centuries (Grinnell 1913; Sheldon 1939), others have suggested that food limitations did not provide a compelling explanation for the historical decline of the species (A. Miller et al. 1965; McMillan 1968), and Snyder and Snyder (2000) found no evidence for food limitations in the remnant population in the 1980s. Koford (1953, 72), one of the few authorities on condors and changes to their historical food supply in southern California, stated:

> With the exception of the killing and molesting of condors by man, the change in the supply of food has been the most important factor determining the distribution and numbers of condors within the last century.

However, Koford did not extend this rationale to explain the disappearance of condors from the Pacific Northwest.

> No evidence has been discovered which suggests that condors were driven from Oregon or northwestern California by a shortage of food there. It is probable that some factor other than lack of food caused them to disappear from these areas. (67)

Koford's analysis of the issue of food supply revolved almost entirely around changes in food availability in southern California, with no details regarding changes in food resources in the Pacific Northwest. Here, we specifically address changes in food resources in the Pacific Northwest and whether or not particular food items may have been increasing or decreasing during the nineteenth century, when condors disappeared from the region.

Salmon

The importance of salmon as a dietary component of condors in the Pacific Northwest is unclear. We know that there were millions of chinook salmon (*Oncorhynchus tshawytscha*), sockeye salmon (*O. nerka*), coho salmon (*O. kisutch*), and steelhead trout (*O. mykiss*) in the Columbia River, Fraser River, Sacramento River, and coastal rivers of northern California and Oregon in the 1800s (Chapman 1986; Northcote and Atagi 1997; Yoshiyama et al. 1998; Meengs and Lackey 2005). Because most species of *Oncorhynchus* die after spawning, a large spawning run would, at least seasonally, produce a significant number of carcasses (Cederholm et al. 1999). Moreover, salmon attempting to swim upstream to spawn occasionally become stranded around waterfalls and barriers as they attempt to leap these barriers. Finally, Indian fishing villages that relied on salmon for sustenance (figure 26) would have produced a significant amount of concentrated fish offal as they cleaned the salmon and hung them out to dry. As John Kirk Townsend noted in a letter to John James Audubon in the 1830s:

> The Californian Vulture inhabits the region of the Columbia River, to the distance of five hundred miles from its mouth,[7] and is most abundant in spring, at which season it feeds on the dead salmon that are thrown upon the shores in great numbers. It is also often

---

7   Townsend (1848, 265) stated, "In my journey across the Rocky Mountains to the Oregon [Territory] in 1834, I kept a sharp look-out for this rare and interesting bird [the California Condor] in all situations on the route, which I thought likely to afford it a congenial dwelling place; but not one did I see. It was not indeed until my return to the coast in the spring from the Sandwich Islands [Hawaiian Islands], where I spent the winter, that I was gratified by a sight of the great Vulture." Upon returning from Hawaii, Townsend traveled as far east as the Blue Mountains, only about 350 miles from the mouth of the Columbia (but makes no specific mention of seeing condors in the Blue Mountains). It is possible that the distance of five hundred miles was based on David Douglas's account that they were observed east of the Cascades up to 49°N latitude (what would become the Canadian border).

Fig. 26. Indians fishing at Celilo Falls, Columbia River, 1899. Photo by Benjamin Gifford. Courtesy of the Oregon Historical Society.

met with near the Indian villages, being attracted by the offal of the fish thrown around the habitations. . . . On the upper waters of the Columbia the fish intended for winter store are usually deposited in huts made of the branches of trees interlaced. I have frequently seen the Ravens attempt to effect a lodgement in these deposits, but have never known the Vulture to be engaged in this way, although these birds were numerous in the immediate vicinity. (Audubon 1840, 12)

In a subsequent note to Audubon, Townsend continues:

[The Californian Vulture] is particularly attached to the vicinity of cascades and falls, being attracted by the dead salmon which strew the shores in such places. The salmon, in their attempts to leap over the obstruction, become exhausted, and are cast up on the beaches in great numbers. Thither, therefore, resort all the unclean birds of the country, such as the present species, the Turkey-Buzzard, and the Raven. . . . Their food while on the Columbia is fish almost exclusively, as in the neighbourhood of the rapids and falls it is always in abundance; they also, like other Vultures, feed on dead animals. (13)

Townsend also mentioned seeing condors "constantly" in areas where salmon were "cast upon the shores" and observed a condor alighting on a salmon that had had become stranded after trying to leap over Willamette Falls, near current-day Oregon City, Oregon (1848). Townsend's are the only firsthand historical accounts we are aware of that mention condors feeding on salmon,[8] although the Corps of Discovery observed them feeding on unidentified fish that had been "thrown up by the waves on the Sea Coast" near Fort Clatsop (Lewis et al. 2002). Peale (1848, 58) stated that condors were "much more numerous in California, from the fact that the carcasses of large animals are more abundant, which they certainly prefer to the dead fish on which they are obliged to feed in Oregon, and all the countries north of the Spanish settlements in California." However, Peale makes no specific mention of observations of condors feeding on salmon. Macoun (1903, 219) reviewed the history of condor observations in British Columbia and described them as "a rare visitant at the mouth of the Fraser River, B.C., apparently attracted by the dead salmon," but no details were provided that can ascertain whether Macoun actually observed them feeding on salmon. Cassin (1856a) described them as occurring in the vicinity of rivers, "living principally on dead fishes." In a separate publication of the same year in *The United States Magazine*, Cassin (1856b, 24) stated:

> The [California Vulture] is frequently seen on the rivers during the fishing season, particularly in the period at which the salmon ascend the streams of fresh water. Many are killed in attempting to pass rapids, and afford food for this Vulture, which, like its smaller relatives, possesses by no means a fastidious appetite. It devours all descriptions of animal refuse, following a deer wounded by the hunter until it sinks in death, or is satisfied with the rejected parts of slaugtered [*sic*] cattle, fresh or in any state of putridity.

However, Cassin did not visit the western United States, and his secondhand account was likely derived from communications with his contemporary and associate at the Philadelphia Academy of Natural Sciences in the 1840s—John Kirk Townsend. Cassin (1856a) also relayed a report

---

8 Thomas Nuttall, a botanist and traveling companion of John Kirk Townsend, also noted condors feeding on salmon, "which [the condors] find wrecked and stunned to death in their unceasing attempts to ascend the rapids of the Oregon [Columbia River] and Wahlamet [Willamette River]" (1840). However, because Townsend and Nuttall were traveling together, we do not consider these to be independent observations, and it is likely that Nuttall was simply repeating what he had heard from Townsend or what he had read from David Douglas (1829).

of a condor that was allegedly dissected by A. S. Taylor, who described the stomach contents as containing "fish, meat, and muscles [*sic*] with the shells on—the latter in a half-digested condition."

The paucity of the historical record regarding condors feeding on salmon might be an artifact of the fragmentary nature of historical observations in the region, or it may be that salmon were consumed infrequently or seasonally. Given the accounts of Townsend, Macoun, and the Corps of Discovery that indicated fish were the primary food resource for condors along the Columbia and Fraser Rivers, we cannot rule out the possibility that salmon carcasses around falls and cascades, spawning areas, and Indian villages may have provided an important food resource for condors in the Northwest. Because there were multiple salmon runs in many rivers that occurred at different times of the year, this food resource was not limited to a single season, although it was less abundant in the mid-Columbia River from November to March. In late April or May the spring chinook salmon run peaked in the Columbia River Gorge just before the annual spring freshet on the Columbia reached its peak (Schoning et al. 1951). This was followed by the peak of the sockeye salmon run in July. Chinook salmon were present throughout the summer and became extremely abundant in early September during the fall run, at which time steelhead and silver salmon were also present. Steelhead were also present in significant numbers from July to October (Schoning et al. 1951).

Contemporary observations of condors eating fish are rare, but they readily consume fish provided in captivity. Other Cathartid vultures appear to be quite adaptable to local foraging conditions, in some situations focusing foraging efforts in areas where humans discard undesirable or injured fish, and even taking live fish under certain circumstances (e.g., Jackson et al. 1978). Among the Karok Tribe, vultures (it is unclear whether they were California Condors or Turkey Vultures) were also known to appear along the Klamath River at the site of smoke that signaled the First Salmon Festival because, as one informant put it, "They know they are going to eat salmon" (Kroeber and Gifford 1949).

At the time that condor observations were in noticeable decline in the region (mid-1800s), salmon runs may have been larger than at just about any other time in postglacial history because Native Americans had experienced massive population declines from introduced diseases and were no longer harvesting large quantities of fish (reviewed by Meengs and

Lackey 2005). Local declines in salmon populations were quickly noticed in northern California and southwest Oregon in the 1850s following the development of hydraulic mining, where miners used pressurized water to blast away hillsides, washing excess sediment into streams and rivers and suffocating spawning fish and their offspring (Schaeffle 1915; Meengs and Lackey 2005). Around the same time, intensive commercial fishing for Pacific salmon was taking hold with the development of an effective method of canning fish (Cobb 1921; Lichatowich et al. 1999). However, in northern California, more than ten million pounds of salmon were still being harvested from the Sacramento River in 1880 (Porter et al. 1882), and the maximum yield of chinook salmon in the Columbia River was not attained until around 1880 (McKernan et al. 1950), when there were still several million adult salmon returning to the Columbia and other Northwest rivers each year (Chapman 1986; Yoshiyama et al. 1998; Meengs and Lackey 2005). In the Fraser River system of British Columbia, it is estimated that there were more than fifty million salmon in the late 1800s and early 1900s, prior to a series of landslides in 1913 caused by railroad construction, which blocked salmon from the entire upper basin (Northcote and Atagi 1997). Thus, although localized salmon declines began in the 1850s, salmon remained relatively abundant in northern California, Oregon, Washington, and British Columbia until the 1880s, when condor observations had become rare in the region, arguing against the hypothesis that declines in salmon populations caused condor declines in the region.

Whales

Beached whales represent a potentially substantial food resource for condors; however, documented historical observations of condors feeding on whale carcasses are rare. We are aware of only three historical accounts of condors feeding on whales:

1. A condor was killed on a Cape Disappointment, Washington, beach next to a whale carcass by the Corps of Discovery on 18 November 1805 (Lewis et al. 2002).[9]

---

9   A bronze statue of a California Condor perched on whale bones now commemorates that observation. It is located at the southern end of the Discovery Trail at the Port of Ilwaco, Washington.

Fig. 27. California Condors feeding on a gray whale near Big Sur, California, 2006. Photo by Ryan Choi. Courtesy of Ventana Wildlife Society.

2. Gambel (1847, 25) reported: "It is not uncommon to see them assemble with the gulls, and greedily devour the carcase of a whale which had been cast ashore," but it is unclear whether this was based on a first-hand observation.

3. A. Taylor (1859) observed "a number of" condors feeding on a whale carcass in 1859 near Monterey, California.

More recently, in 2006 and 2008, condors were documented feeding on gray whale (*Eschrichtius robustus*) carcasses near Big Sur, California (J. Burnett, Ventana Wildlife Society, pers. comm., 2011; figure 27).

Harry Harris (1941, 3–4) gave a colorful, embellished account of the first European encounter with condors, as they feasted on a whale carcass.

The record [of condor observations] begins with the published diary of a barefoot Carmelite friar, Fr. Antonio de la Ascension, who in 1602, from the tossing deck of a tiny Spanish ship, observed on a California beach the stranded carcass of a huge whale (conceivably and probably)[10] surrounded by a cloud of ravenous condors. Here indeed is material with which to stir the most dormant imagination; civilized man for the first time beholding the greatest volant bird recorded in human history, and not merely an isolated individual or two, but an immense swarm rending at their food, shuffling about in crowds for a place at the gorge, fighting and slapping with their great wings at their fellows, pushing, tugging at red meat, silently making a great commotion, and in the end stalking drunkenly to a distance with crop too heavy to carry aloft, leaving space for others of the circling throng to descend to the feast!

While Ascension did see a whale carcass at Monterey Bay (being fed upon by bears), he never reported condors feeding on a whale carcass—something that Harris simply thought was conceivable and probable.

The documented evidence of condors feeding on whales is sparse; however, isotopic data from Pleistocene condor bones suggest that marine resources were an important component of their diet in Pacific coastal regions (Chamberlain et al. 2005; Fox-Dobbs et al. 2006). Commercial whaling practices caused large population declines for many whale species, leading some to suggest that this may have caused or contributed to California Condor population declines in the nineteenth century (C. Smith 2006). Below, we review Native American and commercial whaling practices in the region to see how they might have influenced whale populations and the historical availability of whale carcasses for condors.

Several Pacific Northwest tribes practiced whaling, a hunting tradition that appears to have been limited to tribes on the west side of the Olympic Peninsula, including the Quileute, Quinault, and Makah Tribes, as well as tribes on the west coast of Vancouver Island (Curtis 1915; Waterman 1920; Pettitt 1950; Monks 2001). Although other coastal tribes in the Pacific Northwest south of the Olympic Peninsula did not typically kill whales at sea (Waterman 1920), they would harvest the blubber of beached whales for food and oil (Meany 1907; Waterman 1920; Kroeber 1925). According to Kroeber (1925, 84), among the Yurok Tribe of northern California,

---

10 Harris notes that he was using his imagination here, not reporting a fact.

the stranding of a whale was always a great occasion, sometimes productive of quarrels. The Yurok prized its flesh above all other food, and carried dried slabs of the meat inland, but never attempted to hunt the animal.

Whale strandings, although sporadic, were likely a more common event prior to stock depletions. Mrs. Sarah Ruhamah De Bell observed twenty large whales on one stretch of beach at Clatsop, Oregon, in 1840.

Some of these monsters took a long time to die, but all of them furnished food for the Indians. The white folks saved as much as they could of the oil. (Meany 1907, 14)

Among the Makah Indians, who were some of the most adept native whalers, whale blubber was highly valued for the oil it contained and for its ability to be dried and stored for food (Waterman 1920). The flesh was also taken for food; ligaments were prepared and made into ropes, cords, and bowstrings; and the stomach and intestines were dried and used to hold oil (Swan 1870; Curtis 1913). However, on some occasions whales killed at sea could take one to two days to bring to land for butchering. During that time the intestines, stomach, and flesh would begin to rot, leaving food for scavengers. As Waterman (1920, 46) noted:

The process of decay goes on much more rapidly in the flesh than in the blubber, which keeps for an indefinite period, even if not removed from the whale. The flesh can be removed from the bones only after stripping off the blubber, which requires time. Possibly that is the reason the flesh of the whale is not more generally utilized. Blubber which has become rancid, through overmuch delay, is tried out,[11] and the oil is used for various technological purposes, not for food. The bones, with the muscles and ligaments, are left on the beach for the birds and other scavengers. All of the blubber, however, down to that on the flukes, is carefully preserved.

---

11 "Trying out" is a technical term for extracting oil from the fat of an animal through boiling.

Gray whales and humpback whales (*Megaptera novaeangliae*) were the preferred quarry of the whaling tribes, although other whale species were taken when available (Swan 1870; Monks 2001). Precise records of the number of whales historically harvested by the tribes of the Olympic Peninsula and Vancouver Island are not available. Nicolay (1846) reported that about twenty whales were killed annually in the Strait of Juan de Fuca, with most of the whales appearing in the North Pacific from May to November. The timing of the harvest is consistent with our current knowledge of the timing of the gray whale migration through the region (Sumich 1984) and the movements of humpback whales to nearshore coastal feeding grounds (National Marine Fisheries Service 1991; Clapham et al. 1997; Calambokidis et al. 2001). By the 1860s, the number of whales killed by the Makah had declined (Swan 1870), but it is not clear whether the decline in harvest was due to the availability of whales or a shift to harvesting northern fur seals (*Callorhinus ursinus*), which were easier to take and whose skins and oil could be traded with the Euro-American settlers (Swan 1870; *Puget Sound Argus* 1882). Despite this lack of clarity, we do know that the decline in Native American whale harvest occurred at the same time gray whales were in serious decline in the eastern Pacific due to intensive commercial harvest.

The first commercial whaling vessel reached the Pacific (off the coast of Chile) in 1791 (Tower 1907). By 1808, whalers were hunting sperm whales along the coast of Baja California, and by 1834 there were so many whalers off the coast of California that the Hudson's Bay Company considered opening another trading post to provision them (Busch 1998). In 1838, the great Northwest coast whaling grounds were discovered and by 1850, almost all the Atlantic whale fleets had moved to harvesting whales in the Pacific (Tower 1907; Tønnessen and Johnsen 1982). In 1851, shore whaling was first tried at Monterey, California, and over the course of the next twenty years this practice was carried out from Monterey to Baja California (Tower 1907). Particularly devastating to the gray whale population was the practice of killing whales in their breeding lagoons in Baja California, starting in earnest in 1858 (Tønnessen and Johnsen 1982; Busch 1998). The rapid depletion of stocks soon followed (between 1845 and 1874 over eight thousand gray whales were killed [Busch 1998]), and some believed the gray whale was headed toward extinction (Tønnessen and Johnsen 1982; Busch 1998).[12] As the whales along the California coast were depleted, whalers moved north to Oregon, Washington, and British Columbia, with nearshore commercial whale harvesting continuing well into the twentieth century. Unlike native whaling practices, which sometimes resulted in rotten portions of the carcass being left on the beach, commercial whalers rendered all fleshy parts of the carcass (Webb 1998), leaving nothing for the scavengers. This shift from a limited Native American harvest that left rotting meat on beached carcasses to industrial whaling practices that decimated the whale population occurred around the same time as condor observations in the region were in decline. However, the importance of whales in the condor's historical diet is unclear, as there are only a few known observations of condors feeding on whales.

Seals and Sea Lions

Pinnipeds along the coast of northern California, Oregon, Washington, and British Columbia (most notably harbor seals [*Phoca vitulina*], northern fur seals, California sea lions [*Zalophus californianus*], and Steller's sea lions [*Eumetopia jubatus*]) may once have provided a substantial coastal food resource for condors. Pinnipeds may have also provided a minor food resource for condors farther inland, as some occurred in the major Pacific Northwest river systems (Paulbitski 1974; Lyman et al. 2002) and were occasionally taken by Native American nets and traps (Lyman et al. 2002). Although historical accounts of condors feeding on pinnipeds are uncommon, condors released in central California are regularly observed feeding on beached carcasses of California sea lions at one of their haul-out sites near Big Sur, California (J. Burnett, Ventana Wildlife Society, pers. comm., 2011). In one case a condor was even observed killing a weak and abandoned sea lion pup (M. Tyner, Ventana Wildlife Society, pers. comm., 2011). Below we review observations of seal and sea lion abundance and practices of harvesting pinnipeds to see if shifts in relative abundance and harvest practices were coincident with the condor's range contraction in the nineteenth century.

---

12   The International Whaling Commission has prohibited the killing of gray whales in the Pacific Ocean since 1947 and the gray whale was protected under the Endangered Species Act from 1973 to 1994, when it was one of the first species to be removed from the list due to recovery. The eastern North Pacific gray whale has now recovered to a level near its estimated pre-exploitation size and is considered one of the world's great conservation success stories (Gerber et al. 2000).

Native American tribes in the Pacific Northwest killed seals and sea lions, usually by sneaking up on them at rookeries or haul-out sites, although specific hunting practices varied regionally (Braje and Rick 2011). Prehistoric impacts to mainland rookeries may have been significant, especially those situated on accessible shorelines (Hildebrandt and Jones 1992). However, accounts of explorers and settlers suggest that sea lions and seals were relatively abundant throughout the coastal Pacific Northwest through the mid-1800s (Newberry 1857; Chase 1869; Swan 1870; Murphy 1879), where, except for limited Native American harvest, they remained largely unmolested (Murphy 1879). Significant commercial harvest of pinnipeds, particularly California sea lions and northern fur seals, began in the 1850s, when thousands of individuals were killed for their skins, whiskers, testicles, and blubber (Scammon 1874; Cass 1985). In 1880, over six thousand fur seals were harvested off the northwest coast of the United States (Swan 1887), and over twelve thousand were harvested in Oregon and Washington in 1892 (Wilcox 1895).

Around the turn of the century, government agents began to kill seals and sea lions in Oregon at the mouth of the Columbia River and at their breeding grounds at the mouth of Elk Creek in hopes of protecting dwindling salmon runs (Oregon Fish and Game Commissioner's Board 1901). During the spring and summer of 1900, government agents killed 288 seals and 670 sea lions (Oregon Fish and Game Commissioner's Board 1901). They were also being slaughtered by recreational shooters who viewed them as a nuisance species because of their predatory habits on salmon (Wild 1898). These practices escalated through the early 1900s, greatly depleting seal and sea lion populations along the Oregon coast (Scheffer 1928), but these killings would have also provided a temporary abundance of carcasses for scavengers during the time period when condors disappeared from the Pacific Northwest.[13]

Native Ungulates

The principal native ungulates that were available to condors in the Pacific Northwest included Rocky Mountain elk (*Cervus canadensis nelsoni*), Roosevelt elk (*C. c. roosevelti*), tule elk (*C. c. nannodes*), black-tailed deer (*Odocoileus hemionus columbianus*), and white-tailed deer (*O. virginianus*).

---

13 This also raises the possibility that a significant number of lead-contaminated carcasses were available to condors during this time—see "Lead Poisoning" section, above.

Bison (*Bison bison*) once occurred in some areas of the Pacific Northwest (east of the Cascades) but were scarce by the time early Euro-American explorers entered the region (Wilkes 1849; Kingston 2010).

In northern California, tule and Roosevelt elk were abundant in the early 1800s, but populations declined with the discovery of beaver (*Castor canadensis*) in the northern Sacramento Valley in 1826–1827 by Jedediah Smith and party (McCullough 1969). This discovery led to Hudson's Bay Company fur brigades frequenting these grounds and harvesting large numbers of animals for food up until the mid-1840s (Harper et al. 1967; McCullough 1969; Phillips 1976). In the 1830s, Peale (1848) noted an abundance of large ungulate carcasses in northern California and suggested this was the reason condors were more prevalent in northern California than in Oregon. Sage (1846) also noted that game was plentiful in northern California in the early 1840s, especially in the vicinity of the Tulare and Sacramento Rivers. Newberry (1857) found elk to be common in the valleys of northern California, but not as common as in the early 1800s, when herds of game in the Central Valley rivaled the bison herds on the Great Plains and the antelope herds of South Africa. Despite apparent reductions from historical numbers and the extirpation of tule elk from northern California shortly after the Gold Rush (McCullough 1969), deer and Roosevelt elk were still considered plentiful in many areas of northern California up until the 1870s (Doney et al. 1916).[14]

In Oregon and Washington, elk and deer were widely distributed and relatively abundant during the early Euro-American settlement period, especially west of the Cascade crest (Wilkes 1849; Victor 1872; Merriam 1897; Bailey 1936; Dixon and Lyman 1996; Harpole and Lyman 1999; Lyman 2006). Game populations in the region were likely increasing at the time as a result of decreased hunting pressure from Native Americans (Martin and Szuter 1999; Lyman and Wolverton 2002; Kay 2007), who were in the midst of massive population declines resulting from a series of fatal epidemics. In 1826, David Douglas found elk to be "plentiful in all the woody parts of the country" and "particularly abundant near the coast" (1914, 155).

---

14 In contrast to Roosevelt elk (of coastal northern California, Oregon, and Washington), tule elk (historically inhabiting the Central Valley and central Coast Range of California) were driven close to extinction by the 1850s due to market hunting as well as the spread of agriculture and the concomitant destruction of elk habitat (McCullough 1969).

Significant impacts to native ungulate numbers began during the Gold Rush era, when many mining camps were established in northern California and southwestern Oregon. These mining camps depleted local deer and elk populations, as hunting was unregulated and occurred year-round to supplement what the mining camps could purchase from nearby farmers (Longhurst et al. 1952; Mark 2006). Ranching and grazing operations were also established, and as early as 1851, cattle began replacing elk in some areas (Harper et al. 1967; J. Sawyer 2006). Elk and deer populations were declining in the 1860s around settlements and mining camps, but these ungulates were still considered common in areas that were lightly populated or inaccessible (Cronise 1868). However, from 1870 to 1900, market hunting for deer and elk hides seriously depleted wild ungulate numbers throughout the region as roving commercial hunting outfits penetrated more remote areas (Doney et al. 1916; Hessburg and Agee 2003). In 1880, thirty-five thousand deer hides were shipped from Siskiyou, Trinity, and Shasta Counties in California, with the meat left to rot on the ground (Doney et al. 1916). Similar slaughter was also occurring in Oregon, with thousands of deer hides shipped from southern Oregon to San Francisco in the 1870s and 1880s (*Willamette Farmer* 1887; Doney et al. 1916). Increasing competition from livestock, a series of hard winters from 1879 to 1907, logging practices, and development also contributed to native ungulate declines in the late 1800s (Longhurst et al. 1952).

At the turn of the twentieth century, ungulate numbers were severely depleted (Barnes 1925; Longhurst et al. 1952). Some herds persisted in areas away from settlements, albeit at reduced numbers and distribution (Merriam 1897; Bailey 1936; Longhurst et al. 1952). Deer populations did not begin to stabilize and increase until the early 1900s, following the establishment of hunting seasons, bag limits, and a ban on selling wild game (W. Taylor 1916; Bailey 1936; Longhurst et al. 1952). Elk populations were slower to recover and began to return to significant numbers only in the 1950s and 1960s (Harper et al. 1967; McCullough 1969).

While hunting, hard winters, and other factors ultimately resulted in severe depletion of deer and elk numbers, the quantity of carcasses left in the field for scavengers during this period would have been locally significant, albeit ephemeral (e.g., Doney et al. 1916).[15] Furthermore, in areas

---

15 This also raises the possibility that a significant number of lead-contaminated carcasses were available to condors during this time—see "Lead Poisoning" section, above.

around settlements, where ungulates became scarce in the late 1800s, they were being replaced on the landscape with livestock, which would have provided an alternative, more predictable and reliable food resource for condors.

Livestock

In California, early settlers maintained a few cattle ranches in the northern part of the state during the early 1800s, but at that time, the cattle industry was run largely by Spanish missions and the hide-and-tallow trade of southern California. Only about 3,700 sheep and 55,000 cattle resided in northern California in 1850 (US Department of the Interior 1853; Burcham 1957). Following the Gold Rush, demand for beef instigated enormous cattle drives from southern to northern California to feed the burgeoning mining towns along the Sierra foothills (Cleland 1951; Burcham 1957). Some southern ranchers leased grazing rights in the vicinity of San Jose, Sacramento, or San Francisco Bay, where the stock was fattened after the long journey north (Cleland 1951). Tens of thousands of cattle were also imported from the Midwest via overland trails around this time due to the high price of beef in northern California (Cleland 1951; Burcham 1957). Cattle numbers peaked in California around 1860, when there were 1.1 million head, then declined to about 800,000 in 1880 following a succession of droughts that devastated the cattle industry (Porter et al. 1882; Burcham 1957). However, cattle numbers recovered to more than 1.3 million by 1890 (US Department of the Interior 1895; Burcham 1957). Cattle numbers were much higher in southern and central California through the late 1800s, but there were over 100,000 cattle in northern California by 1860 and more than 175,000 cattle there by 1880 (Burcham 1957).

Sheep populations also increased rapidly in California following the first wave of the Gold Rush. Spaniards first introduced them to California in 1773, but there were only about seventeen thousand sheep in California in 1850 (Burcham 1957). By 1860 there were more than one million sheep in California, a number that more than doubled by 1870 and surpassed four million by 1880 (Burcham 1957). Although a large portion of the sheep were in central and southern California prior to 1900, there were over eight hundred thousand sheep in northern California by 1880 (Burcham 1957).

Fig. 28. A large flock of sheep in the Pacific Northwest, circa 1902. Photo by Benjamin Gifford. Courtesy of the Oregon Historical Society, negative Gi 221.

In the Oregon Territory, livestock introductions began with the importation of hogs and sheep by the Astor party in 1811 (Carey 1922), and the Hudson's Bay Company imported cattle from California starting in 1830 (Gaston 1912). The initial quantity of livestock was modest and largely meant for the subsistence of settlers, who kept small herds of dairy cows, goats, oxen, sheep, horses, and beef cattle on family farms (Hessburg and Agee 2003). In 1841 there were approximately three to ten thousand cattle in the Willamette Valley, five hundred horses, and a "multitude of hogs" (Bancroft 1886). The mass influx of immigrants following the Gold Rush in 1849 brought large numbers of livestock to the region (Carey 1922), and cattle and sheep herds expanded well beyond family farming into an industry (figure 28). In Oregon, by the 1860 census, there were approximately 147,000 cattle and over 1.6 million sheep (US Department of the Interior 1864); by 1890 there were over 520,000 cattle and 1.7 million sheep (US Department of the Interior 1895); and by 1900 there were 532,000 cattle and approximately 2.8 million sheep (US Department of the Interior 1902).

Livestock numbers also increased in Washington later in the nineteenth century, although not in the numbers seen in Oregon and California. In 1860, there were only 25,888 cattle and 10,157 sheep in the Washington Territory (US Department of the Interior 1864). At the time of the 1890 agricultural census, there were 255,134 cattle and 265,267 sheep (US Department of the Interior 1895), and by 1900, there were 290,000 cattle and 864,480 sheep (US Department of the Interior 1902).

Although most of the livestock industry in Oregon and Washington was situated east of the Cascades in Oregon (and therefore east of most of the known condor occurrence records), there were still large numbers of livestock being imported to the western portions of the two states, especially Oregon. According to the 1890 census there were over two hundred thousand cattle and three hundred thousand sheep in western Oregon and approximately seventy-five thousand cattle and thirty thousand sheep in western Washington (US Department of the Interior 1895).

Early livestock drives sometimes resulted in large losses, which would have meant increased food availability for condors. For example, after buying eight hundred head of cattle in San Francisco and San Jose in 1837, Ewing Young and others drove the herd north to Oregon (Dillon 1961). By the time they reached the Willamette settlements in western Oregon, they had lost 168 cattle (Dillon 1961).

The number of domestic livestock increased rapidly in the Pacific Northwest following the California Gold Rush, with millions of animals present by the late 1800s. Even if only a small fraction of these animals died and were accessible to condors annually, there would have been thousands of carcasses to feed on. In reality, mortality, especially among sheep, was sometimes extremely high. For example, the 1890 census reported five hundred thousand sheep in Oregon dying from disease and weather in 1889 (a mortality rate of approximately 20 percent; US Department of the Interior 1895), with more than thirty-one thousand dying in western Oregon alone (a mortality rate of approximately 10 percent).

Declines in condor observations were noted following introduction and proliferation of livestock to the region. An exponentially expanding livestock industry at the time of condor population declines is inconsistent with the hypothesis that food limitations were the cause of the decline.

Summary of Changes to the Condor's Food Supply

Our review of the historical record indicates that the lack of food does not appear to have caused the condor's nineteenth-century population decline in the Pacific Northwest. There were diverse and abundant native food resources in the region through the 1850s, including salmon, whales, pinnipeds, and ungulates (figure 29). Although declines in some native food resources were observed in the region in the mid to late 1800s, it is unlikely food was becoming limited for condors in the Pacific Northwest for a number of reasons:

1. Obligate scavengers have evolved mechanisms to cope with cyclic or temporary food shortages. As large soaring birds with keen eyesight, condors are adapted to searching enormous areas for food while expending little effort. With the ability to travel long distances and soar at high altitudes, condors can search several hundred square kilometers for food on days when soaring conditions are favorable. Furthermore, they have an extremely flexible diet, which includes fish, small mammals, livestock, native ungulates, and marine mammals. Their ability to switch between food resources and soar long distances are effective adaptations to exploiting seasonal, migratory, or temporary food resources and an effective buffer against local or temporary food shortages.

2. Hunting practices resulted in temporary increases in carcass availability in the mid to late 1800s. Market hunters were shooting thousands of elk and deer for their hides in northern California and southern Oregon in the late 1800s and leaving the meat and viscera in the field. Seals and sea lion carcasses would also have been abundant from commercial harvest operations. Furthermore, the lack of regulated hunting seasons in the 1800s meant that carcasses would have been available throughout the year. While this ultimately resulted in lower numbers of ungulates and pinnipeds at the turn of the twentieth century, for a number of years there would have been a superabundance of carcasses for condors and other scavengers. Many of these carcasses may have been contaminated with lead bullet fragments, which could have played a role in the condor's decline (see "Lead Poisoning" section, above).

3. Around the time that native ungulates, whales, and pinnipeds were in decline in the region, there were massive increases in livestock, numbering hundreds of thousands of individuals, many of which roamed

freely in the mountain ranges prior to grazing restrictions. Mortality rates of these livestock herds were relatively high, sometimes exceeding 20 percent, due to the inferior animal husbandry practices at the time.

Summary of the Plausible Causes of Extinction in the Pacific Northwest

Identifying and eliminating, or sufficiently reducing, the original causes of local extinction are critical to successful reintroduction planning (IUCN 1987; B. Griffith et al. 1989). However, extinction processes are often the result of interactions between multiple causes that are difficult to disentangle (Carrete et al. 2007). This can be especially problematic when working with data sets assembled from historical observations that have inextricable statistical biases. Despite these problems, a careful review of the range of plausible hypotheses can be helpful in rejecting those that do not align

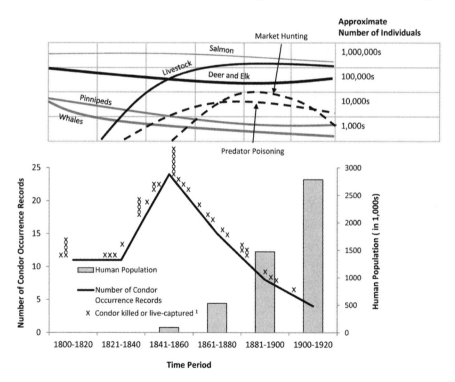

[1] Excludes record #40 due to uncertain time period (1850-1880) (See table 3 and appendix for details).

Fig. 29. Relative changes in California Condor food resources and associated threats (top panel) in relation to the number of California Condor occurrence records, human population, and numbers of condors killed or captured and removed from the wild (bottom panel) in the Pacific Northwest, 1800–1920.

with the scope, magnitude, or timing of the decline of the subject species. It can also identify areas where additional research may help reject other hypotheses. Contemporary information regarding the causes of decline in ecological surrogates can also be helpful in such reviews.

Based on our review, there is limited direct observational evidence—but significant circumstantial evidence—that secondary poisoning could have been a primary factor in the condor's extirpation from the Pacific Northwest (table 5). Strychnine-laced carcasses were locally abundant near homesteads, trapping sites, and travel corridors during the period of the condor's population decline, and the social feeding habits of condors predispose them to mass die-offs when their food is contaminated. Moreover, contemporary studies show that predator poisoning is a major factor in population declines for many vulture species around the world. Lead poisoning may also have played a role in the condor's range contraction, but there is uncertainty regarding the bioavailability of lead to condors during this time period due to changes in firearms and ballistics. Collecting and shooting resulted in some direct loss to the condor population and may also have played a role in the condor's range contraction. Loss of nesting habitat, egg collecting, and Native American ritual sacrifice are not likely causes of the condor's disappearance from the region.

Although declines in some native food resources, particularly coastal resources, were occurring around the time that condor occurrence records were also in decline, it is unlikely that food was ever limiting for condors in the Pacific Northwest. Market hunting ultimately reduced native ungulate and pinniped populations in the late 1800s but also temporarily produced a significant amount of offal in some areas. Furthermore, there was a massive importation of livestock to the Pacific Northwest by the time these native food resources were significantly depleted.

It should not be surprising that anthropogenic causes were almost certainly responsible for the decline and extirpation of condors from the Pacific Northwest. Human influence is highly correlated with range contractions for many North American species (Laliberte and Ripple 2004). But the question remains: Why did populations go extinct in the Pacific Northwest while they persisted in southern California (albeit in a rapidly declining state)? Is it possible that the habitat conditions in the Northwest were marginal, contributing to their early demise compared with the core range in southern California?

Table 5. Evaluation of extinction hypotheses for California Condors in the Pacific Northwest.

Hypothesis	Correct time frame?	Potential geographic extent and magnitude of effects[a]	Suspected cause of other vulture species' declines?	Reject hypothesis?[b]
Secondary poisoning	Y	Regional/very high	Y	N
Lead poisoning	Y	Regional/very high	Y	N
DDT/DDE and eggshell thinning	N	Regional/unknown	N	Y
Collecting and shooting	Y	Regional/high	Y (locally)	N
Egg collecting	N	Local/low	N	Y
Loss of nesting habitat	N	Local/low	Y (locally)	Y
Native American ritual killings	N	Local/low	N	Y
Food increases in southern and central California	N	Regional/low	N	Y
Food declines	N	Regional/high	Y (locally)	Y

a   See text and figure 29 for details.
b   We rejected those hypotheses that were inconsistent with the time frame of the decline and those whose potential impacts had low magnitude or were localized. Information on whether the hypotheses have been the documented cause of other vulture species' declines is given for reference but was not a factor in rejecting any hypotheses.

It seems intuitive that populations at the periphery of a species' range (such as condors in the Pacific Northwest) might be innately more susceptible to extinction events than those populations at the core of the range. This is because peripheral populations often represent the extremes of the species' ecological niche, where species should be expected to occur in lower densities or lower quality habitats (G. Caughley et al. 1988). However, populations at the edge of a species' range are not always less dense, and in many cases peripheral populations persist despite range contractions (Channell and Lomolino 2000).

Given its extensive historical range, it is likely that the California Condor was once divided among a number of interacting populations. If populations in the Pacific Northwest were indeed on the periphery of the condor's ecological niche, they may have been more susceptible to increases in mortality or declines in productivity than those populations at the core of the species' range in southern California. However, it is also possible that threats were unevenly applied to the landscape (see Rodríguez 2002), whereby the northern populations were subjected to more shooting and poisoning early in the settlement of the West than were condors in the mountains of southern California, and thus they declined and went extinct faster. Both scenarios are plausible, as is the scenario that the asynchronous distribution of threats and the location of populations at the periphery of the range were factors contributing to the extinction of condors in the Pacific Northwest. Thus, it is unclear whether the peripheral position in the species' range was part of the reason condor populations in the Pacific Northwest went extinct first. Nevertheless, peripheral portions of the California Condor's historical range may be important to their recovery, because the location of a reintroduction site with respect to the core or periphery of the species' historical range is not always a good predictor of reintroduction success (White et al. 2012). This is especially true when the reasons for local extinction are anthropogenic causes that have since evaporated or been significantly dampened, or in circumstances (as with the condor) when threat factors have since been completely reshuffled in form and space.

Chapter 5
Summary and Future Outlook

co·ex·ist (kō'ĭg-zĭst') *vi.* To exist together in the same place or at the same time.

> One very important thing can be learned from the record of the past:
> If man chooses to do so, and with no more than token sacrifice, he
> can live with the condor. Whether modern man is capable of making
> this slight sacrifice may well determine his own fitness for survival.
> (McMillan 1968, 177)

The long absence of the California Condor from the Pacific Northwest and
the crisis situation of condor conservation in the late twentieth century
have, until recently, resulted in a lack of focus on recovery efforts in the
Pacific Northwest. It is our hope that by articulating the history of the
California Condor in the region we might inform future dialogue over the
role of the Pacific Northwest in condor conservation.

From our historical review it is clear that the California Condor was
(and still is) culturally important to many Native American tribes in the
region and was regularly observed and collected by early explorers and
settlers. The species' historical occurrence in the region is verified by prehistoric and historic physical evidence and numerous firsthand accounts.
The number and extent of occurrence records demonstrate that human
observations and interactions with condors in the Pacific Northwest were
far more prevalent than previously reported (figure 15).

Evidence now strongly suggests that condors were a resident species in
the region. The long breeding and rearing period of the California Condor
make long-distance seasonal migration extremely unlikely. Furthermore,
two juvenile birds were shot in the Pacific Northwest and preserved in
museums. Because young condors often stay in close proximity to their
nest site for the first year or two of life, it is possible that these birds were

hatched in the region and that at least a portion of the condor population was resident. Additional research into the genetic makeup of individuals from the Pacific Northwest could provide further insights into the historical population structure of the condor.

From a reconstruction of the (admittedly fragmented) historical record, the most plausible hypothesis for the population decline and range collapse of the California Condor is increased mortality from secondary poisoning from laced carcasses and possibly from collecting and indiscriminant shooting. Although direct evidence of condors being poisoned in the Pacific Northwest is limited, the feeding habits of condors make them particularly vulnerable to this threat. Moreover, contemporary studies show that predator poisoning is a major factor in population declines for many vulture species around the world. Collecting and shooting resulted in some direct loss to the condor population, and although the number of reports of condors killed, captured, or removed from the population in the Pacific Northwest is relatively small, there may have been significant losses that went unreported. Lead poisoning is the leading hypothesis for the lack of a current self-sustaining wild condor population (Walters et al. 2010; Finkelstein et al. 2012; Rideout et al. 2012), and it may also have played a role in the condor's range contraction. However, there is some uncertainty regarding the bioavailability of lead to condors during this time period due to differences in firearms and ballistics—which may have resulted in less lead fragmentation—prior to the turn of the century. Loss of food resources, loss of nesting habitat, egg collecting, and Native American ritual sacrifice were not likely causes of the condor's disappearance from the region.

Reintroductions: Challenges and Opportunities

The number of species reintroduction projects has skyrocketed in recent decades (Seddon et al. 2007). This is not surprising, as these projects attract considerable public attention and many zoos have shifted toward a more conservation-oriented mission that lends itself to breeding animals for reintroduction (Seddon et al. 2007). Condor reintroductions are no exception, with a high level of media interest for every egg hatched and public gatherings to witness condor releases to the wild. Although the captive breeding program has secured the near-term survival of the condor, the species' long-term prospects remain uncertain. The primary contemporary threat to the condor's survival and recovery—ingestion of lead from ammunition in carcasses—continues to plague the conservation program at all release sites, despite regulatory and voluntary measures to reduce this threat (Walters et al. 2010; Finkelstein et al. 2012; Rideout et al. 2012). In addition, there are novel threats that have emerged since the range collapse, including (1) collision hazards with overhead wires (Meretsky et al. 2000) and wind turbines (Barrios and Rodríguez 2004),[1] (2) ingestion of microtrash (Mee et al. 2007), and (3) marine contaminants and associated eggshell abnormalities (Kiff et al. 1979; Kiff 1989; but see Snyder and Meretsky 2003). There has also been significant land development and changes in agricultural practices throughout the condor's historical range, but especially in southern California, raising questions about the future prospects of condor recovery in what has traditionally been considered the species' stronghold.

Could the Pacific Northwest offer a safer environment for the California Condor and aid in the species' overall recovery? Perhaps, but some caution is warranted. First, the landscape in the Pacific Northwest has undergone some dramatic changes in the last century, most notably through human population expansion and associated residential, commercial, and agricultural development. Energy development, in the form of wind farms and overhead transmission lines, has also expanded rapidly over the last few decades. A major international airport now sits near historical foraging grounds for condors. Food resources have shifted due to changes in livestock economies, changes in marine mammal densities and distribution, and changes in riverine systems, with the associated reductions in salmon populations ascending waterfalls (where they may have once provided a significant predictable food resource for condors). Species assemblages have changed, with regional apex predator extinctions (e.g., gray wolves and grizzly bears)[2] and changes in the numbers and distribution of avian predators and scavengers. By-products of industry have entered the food

---

1 Although no California Condors have yet been killed by wind turbines, significant mortality of several other vulture species has been documented (e.g., de Lucas et al. 2012), and large soaring raptors may be particularly vulnerable to this threat given the overlap in areas that generate the most wind power and areas that are regularly patrolled by large raptors due to their favorable soaring conditions and open habitats.

2 Recovery efforts are underway in the Pacific Northwest for both species. Grizzly bears remain rare in the conterminous United States outside of the northern Rocky Mountains, but wolves have rapidly expanded into Oregon and Washington in the last few years.

chain and, in some cases, bioaccumulated in potential food items. Finally, fire regimes and forest structure have changed over time, and forests have invaded many historically open mountaintop balds—potentially rendering some historical foraging areas too densely packed with trees for condors to effectively exploit. Therefore, factors that may warrant consideration when planning a future reintroduction include (1) the availability of nesting habitat (including potential tree nests) secluded from areas of high human population density, (2) climatic and atmospheric variables, (3) the amount of food available and accessible to condors, (4) the location of overhead transmission wires, wind turbines, and other collision risks, and (5) potential marine and terrestrial contaminants in food resources.

Conservation programs that involve captive breeding and reintroductions are expensive, time consuming, and usually unsuccessful (Snyder et al. 1996). Nonetheless, the California Condor captive breeding program has been extraordinarily effective, and there are so many captive birds that zoos and release sites will soon reach their capacity (J. McCamman, US Fish and Wildlife Service Condor Coordinator, pers. comm., 2012), meaning there are enough birds to initiate another release site. However, existing release sites continue to have significant operating costs, management challenges, and unsustainable mortality rates due to poisoning from lead ammunition (Walters et al. 2010). Adding yet another release site in the Pacific Northwest without first removing the current primary threat to the species (i.e., lead ammunition in their food) is unlikely to substantively improve the species' status without considerable management intervention (e.g., annually chelating birds that are exposed to lead, monitoring nests, supplemental feeding with clean carcasses, and tracking all birds with radiotelemetry or GPS). Conversely, a wait-and-see approach will mean more habitat degradation and expansion of the human footprint throughout the region without consideration of condor recovery needs. Developing a vision of condor recovery that actively considers the full recent historical range of the condor and outlines necessary steps to recover the species would be a logical next step in identifying those areas in the region that are essential to conserving the species. Whatever path is taken, removing the threats that continue to cause unsustainable mortality will be necessary to achieve species recovery. Without effective protection of those areas that are essential to the long-term viability of a self-sustaining California Condor metapopulation, conservation options will continue to be narrowed.

Establishing appropriate spatial and numerical recovery goals for threatened and endangered species that have been extirpated from much of their former range presents challenges to decision makers, as the question of how much is enough in setting conservation objectives is not a simple matter (Tear et al. 1996; Carroll et al. 2010). However, information on the historical distribution of a species can be helpful in establishing a baseline to facilitate the development of recovery actions that fully consider the effects of anthropogenic changes in the last several hundred years rather than shifting the baseline to the current crisis situation (sensu Pauly 1995).

Initially, condor recovery strategies were focused on the immediate crisis of impending extinction and did not consider the condor's historical range. Despite the successes of the recovery program to date, lack of detailed information on the history of California Condors in the Pacific Northwest has continued to limit exploration of the role this region could play in their recovery. We hope this book will fill this important information gap but acknowledge that additional analyses will be needed to determine current habitat suitability and to map the distribution of novel stressors that were not present when condors soared the Northwest skies over a century ago. These analyses are currently underway by the authors.

Extinction still hangs over the California Condor like the sword of Damocles, held fast only by a thin thread of dedicated professionals and volunteers. Nearly skewered by the blade of extinction in the 1980s, the condor is now on the path to recovery but still exists in a few small populations that continue to persist only because of intensive management intervention. Much has been done, but more is needed. If remaining threats can be controlled or eliminated, restoration of California Condors to the Pacific Northwest could not only restore an important piece of the region's natural heritage but could also establish another hedge against extinction for one of the rarest and most iconic birds in the world.

Literature Cited

Aguilar, G. W., Sr. 2005. When the River Ran Wild: Indian Traditions on the Mid-Columbia and the Warm Springs Reservation. Oregon Historical Society Press: Portland, Oregon.

Alagona, P. S. 2004. Biography of a "feathered pig": The California Condor conservation controversy. Journal of the History of Biology 37:557–83.

Albright, G. L. 1921. Official Explorations for Pacific Railroads. University of California Publications in History, vol. 11. University of California Press: Berkeley.

Alekseev, A. I. 1987. The Odyssey of Russian Scientist: I. G. Voznesenskii in Alaska, California and Siberia 1839–1849. Alaska History 30. Limestone Press: Kingston, Ontario.

Alerstam, T., A. Hedenström, and S. Åkesson. 2003. Long-distance migration: Evolution and determinants. Oikos 103:247–60.

Amadon, D. 1964. The evolution of low reproductive rates in birds. Evolution 18:105–10.

Anthony, A. W. 1893. Birds of San Pedro Martir, lower California. Zoe 4:228–47.

Aoki, H. 1994. Nez Perce Dictionary. University of California Publications in Linguistics, vol. 122. University of California Press: Berkeley.

Arizona Condor Review Team. 2002. A Review of the First Five Years of the California Condor Reintroduction Program in Northern Arizona. Prepared for the California Condor Recovery Team and USFWS California/Nevada Operations Office. 27 February 2002 printing.

Audubon, J. J. 1840. The Birds of America, From Drawings Made in the United States and Their Territories. J. J. Audubon: New York; J. B. Chevalier: Philadelphia.

Avery, B. P. 1874. Ascent of Mount Shasta. Overland Monthly and Out West Magazine 12:466–76.

Bailey, V. 1936. The mammals and life zones of Oregon. North American Fauna 55:1–416.

Baird, S. F., J. Cassin, and G. N. Lawrence. 1858. General Report Upon the Zoology of the Several Pacific Railroad Routes: Birds. Vol. 9, part 2 of Reports of Explorations and Surveys, to Ascertain the Most Practicable and Economical Route for a Railroad From the Mississippi River to the Pacific Ocean. Beverly Tucker, Printer: Washington, DC.

Bancroft, H. H. 1886. History of Oregon, Vol. 1: 1834–1848. History Company Publishers: San Francisco.

Barlow, M. S. 1902. History of the Barlow Road. The Quarterly of the Oregon Historical Society 3:71–81.

Barnes, E. P. 1925. Elk in Del Norte County. California Fish and Game 11:90.

Barnston, G. 1860. Abridged sketch of the life of Mr. David Douglas, botanist, with a few details of his travels and discoveries. Canadian Naturalist and Geologist and Proceedings of the Natural History Society of Montreal 5:120–32, 200–208, 267–78, 329–49.

Barrett, S. A. 1908. The ethno-geography of the Pomo and neighboring Indians. University of California Publications in American Archaeology and Ethnology 6:1–400.

Barrios, L., and A. Rodríguez. 2004. Behavioural and environmental correlates of mortality at on-shore wind turbines. Journal of Applied Ecology 41:72–81.

Bates, C. D. 1983. The California collection of I. G. Voznesenski. American Indian Art Magazine 8:36–41, 79.

Bates, C. D., J. A. Hamber, and M. J. Lee. 1993. The California Condor and California Indians. American Indian Art Magazine 19:40–47.

Baumhoff, M. A. 1958. California Athabascan groups. Anthropological Records 16:157–238.

Bayliss, C. K. 1909. The significance of the Piasa. Transactions of the Illinois State Historical Society 13:114–22.

Bedrosian, B., and D. Craighead. 2009. Blood lead levels of Bald and Golden Eagles sampled during and after hunting seasons in the Greater Yellowstone Ecosystem. Extended abstract *in* R. T. Watson, M. Fuller, M. Pokras, and W. G. Hunt (editors), Ingestion of Lead from Spent Ammunition: Implications for Wildlife and Humans. The Peregrine Fund, Boise, Idaho, USA, DOI 10.4080/ilsa.2009.0209.

Beidleman, R. G. 2006. California's Frontier Naturalists. University of California Press: Berkeley and Los Angeles.

Belding, L., in litt. 1878. Letter from Lyman Belding to Robert Ridgway, 21 March 1878 (copy in Smithsonian Institution Archives, United States National Museum, Division of Birds, records circa 1854–1959, record unit 105).

Belding, L. 1879. A partial list of the birds of central California (edited by R. Ridgway). Proceedings of the National Museum 1:388–449.

Belding, L. 1890. Land Birds of the Pacific District. Occasional Papers of the California Academy of Sciences 2: San Francisco.

Bendire, C. 1892. Life Histories of North American Birds: With Special Reference to Their Breeding Habits and Eggs, with Twelve Lithographic Plates. United States National Museum Special Bulletin 1. Government Printing Office: Washington, DC.

Berg, L. (editor). 2007. The First Oregonians. 2nd ed. Oregon Council for the Humanities: Portland, Oregon.

Bernis, F. 1983. Migration of the common Griffon Vulture in the western Palearctic. Pages 185–96 *in* S. R. Wilbur and J. A. Jackson (editors), Vulture Biology and Management. University of California Press: Berkeley.

Berreman, J. V. 1944. Chetco archaeology: A report of the Lone Ranch Creek shell mound on the coast of southern Oregon. General Series in Anthropology 11:1–40.

Bidwell, J. 1890. Life in California before the gold discovery. Century Magazine 41:163–83.

Bijleveld, M. 1974. Birds of Prey in Europe. Macmillan Press: London.

Bildstein, K. L. 2004. Raptor migration in the Neotropics: Patterns, processes, and consequences. Ornithología Neotropical 15 (Suppl.): 83–99.

Bildstein, K., M. J. Bechard, C. Farmer, and L. Newcomb. 2009. Narrow sea crossing presents major obstacles to migrating Griffon Vultures *Gyps fulvus*. Ibis 151:382–91.

Blake, W. C. 1895. Big price for a bird skin. Nidiologist 2:96, 122.

Bledsoe, A. J. 1885. Indian Wars of the Northwest: A California Sketch. Bacon & Company: San Francisco.

Blomkvist, E. E. 1972. A Russian scientific expedition to California & Alaska 1839–1849, the drawings of I. G. Voznesenski. Oregon Historical Quarterly 73:100–170.

Blueweiss, L., H. Fox, V. Kudzma, D. Nakashima, R. Peters, and S. Sams. 1978. Relationships between body size and some life history parameters. Oecologia 37:257–72.

Boas, F. 1901. Kathlamet Texts. Smithsonian Institution, Bureau of American Ethnology, Bulletin 26. Government Printing Office: Washington, DC.

Boas, F. 1910. Kwakiutl Tales. Vol. 2, Columbia University Contributions to Anthropology. Columbia University Press: New York.

Boas, F. 1916. Tsimshian Mythology, Based on Texts Recorded by Henry W. Tate. Pages 29–1037 *in* Thirty-first Annual Report of the Bureau of American Ethnology to the Secretary of the Smithsonian Institution, 1909–1910. Government Printing Office: Washington, DC.

Borneman, J. C. 1966. Return of a condor. Audubon Magazine 68:154–57.

Boyd, R. 1999. The Coming of the Spirit of Pestilence: Introduced Infectious Diseases and Population Decline Among Northwest Coast Indians, 1774–1874. University of Washington Press: Seattle.

Braje, T. J., and T. C. Rick. 2011. Human Impacts on Seals, Sea Lions, and Sea Otters: Integrating Archaeology and Ecology in the Northeast Pacific. University of California Press: Berkeley.

Brasso, R. L., and S. D. Emslie. 2006. Two new late Pleistocene avifaunas from New Mexico. Condor 108:721–30.

Breschini, G. S., and T. Haversat. 2000. Archaeological data recovery at CA-SCR-44 at the site of the Lakeview Middle School, Watsonville, Santa Cruz County, California. Archives of California Prehistory 49. Coyote Press: Salinas, California.

Broecker, W. S., G. H. Denton, R. L. Edwards, H. Cheng, and R. B. Alley. 2010. Putting the Younger Dryas cold event into context. Quaternary Science Reviews 29:1078–81.

Brooks, A., in litt. 1931. Letter from Allan Brooks to W. Lee Chambers, 6 March 1931. Historical correspondence, Museum of Vertebrate Zoology, University of California, Berkeley.

Broughton, J. M. 2004. Prehistoric human impacts on California birds: Evidence from the Emeryville Shellmound avifauna. Ornithological Monographs 56:1–90.

Broughton, J. M., D. Mullins, and T. Ekker. 2007. Avian resource depression or intertaxonomic variation in bone density? A test with San Francisco Bay avifauna. Journal of Archaeological Science 34:374–91.

Brown, J. H. 1892. Brown's Political History of Oregon, vol. 1. Wiley B. Allen: Portland, Oregon.

Browne, J. H. 1914. The redwood of California. American Forestry 20:795–802.

Bryant, W. E. 1891. Andrew Jackson Grayson. Zoe 2:34–68.

Bulletin (San Francisco newspaper). 1856. The California Condor. 16 May.

Bulletin (San Francisco newspaper). 1864. A pretty pet! 21 December.

Bulletin (San Francisco newspaper). 1868. Proud bird of the mountain. 19 August.

Bulletin (San Francisco newspaper). 1880. State news in brief. 5 April.

Burcham, L. T. 1957. California Range Land: An Historico-Ecological Study of the Range Resource of California. California Department of Natural Resources, Division of Forestry: Sacramento.

Busch, R. H. 1998. Gray Whales: Wandering Giants. Orca Book Publishers: Custer, Washington.

Cade, T. J. 2007. Exposure of California Condors to lead from spent ammunition. Journal of Wildlife Management 71:2125–33.

Cade, T. J., S. A. H. Osborn, W. G. Hunt, and C. P. Woods. 2004. Commentary on released California Condors *Gymnogyps californianus* in Arizona. Pages 11–25 *in* R. D. Chancellor and B.-U. Meyburg (editors), Raptors Worldwide. Proceedings of the 6th World Conference on Birds of Prey and Owls. WWGBP/MME-Birdlife, Hungary.

Calambokidis, J., G. H. Steiger, J. M. Straley, L. M. Herman, S. Cerchio, D. R. Salden, R. J. Urbán, J. K. Jacobsen, O. von Ziegesar, K. C. Balcomb, C. M. Gabriele, M. E. Dahlheim, S. Uchida, G. Ellis, Y. Miyamura, P. L. De Guevara P., M. Yamaguchi, F. Sato, S. A. Mizroch, L. Schlender, K. Rasmussen, J. Barlow, T. J. Quinn II. 2001. Movements and population structure of humpback whales in the North Pacific. Marine Mammal Science 17:769–94.

California Academy of Natural Sciences. 1863. Proceedings of the California Academy of Natural Sciences, vols. 1, 2: 1854–1862. Towne & Bacon, Excelsior Printing Office: San Francisco.

Campbell, K. E., Jr., and A. T. Stenger. 2002. A new teratorn (Aves: Teratornithidae) from the Upper Pleistocene of Oregon, USA. Pages 1–11 in Z. Zhou and F. Zhang (editors), Proceedings of the 5th Symposium of the

Society of Avian Paleontology and Evolution, Beijing, 1–4 June 2000. China Science Press: Beijing.

Campbell, K. E., and E. Tonni. 1981. Preliminary observations on the paleobiology and evolution of Teratorns (Aves: Teratornithidae). Journal of Vertebrate Paleontology 1:265–72.

Campbell, K. E., and E. Tonni. 1983. Size and locomotion in Teratorns (Aves: Teratornithidae). Auk 100:390–403.

Campbell, R. W., N. K. Dawe, I. McTaggart-Cowan, J. M. Cooper, G. W. Kaiser, and M. C. E. McNall. 1990. The Birds of British Columbia, vol. 2. Nonpasserines, Diurnal Birds of Prey Through Woodpeckers. Royal British Columbia Museum: Victoria, British Columbia.

Caras, R. A. 1970. Source of the Thunder: The Biography of a California Condor. University of Nebraska Press: Lincoln.

Carey, C. H. 1922. History of Oregon. Pioneer Historical Publishing Company: Portland, Oregon.

Carey, C. H. 1935. A General History of Oregon Prior to 1861. Metropolitan Press: Portland, Oregon.

Carrete, M., J. M. Grande, J. L. Tella, J. A. Sánchez-Zapata, J. A. Donázar, R. Diaz-Delgado, and A. Romo. 2007. Habitat, human pressure, and social behavior: Partialling out factors affecting large-scale territory extinction in an endangered vulture. Biological Conservation 136:143–54.

Carroll, C., J. A. Vucetich, M. P. Nelson, D. J. Rohlf, and M. K. Phillips. 2010. Geography and recovery under the U.S. Endangered Species Act. Conservation Biology 24:395–403.

Cass, V. L. 1985. Exploitation of California sea lions, *Zalophus californianus*, prior to 1972. Marine Fisheries Review 47:36–38.

Cassin, J. 1856a. Illustrations of the Birds of California, Texas, Oregon, British and Russian America; Intended to Contain Descriptions and Figures of All North American Birds Not Given by Former American Authors, and a General Synopsis of North American Ornithology. J. B. Lippincott & Co.: Philadelphia.

Cassin, J. 1856b. Ornithology of the United States, and British and Russian America; the American vultures. United States Magazine 3:18–29.

Caughley, G., D. Grice, R. Barker, and B. Brown. 1988. The edge of the range. Journal of Animal Ecology 57:771–85.

Caughey, J. W. 1948. The California Goldrush. University of California Press: Berkeley.

Ceballos, O., and J. A. Donázar. 1990. Roost-tree characteristics, food habits and seasonal abundance of roosting Egyptian Vultures in northern Spain. Journal of Raptor Research 24:19–25.

Cederholm, C. J., M. D. Kunze, T. Murota, and A. Sibatani. 1999. Pacific salmon carcasses: Essential contributions of nutrients and energy for aquatic and terrestrial ecosystems. Fisheries 24:6–15.

Chamberlain, C. P., J. R. Waldbauer, K. Fox-Dobbs, S. D. Newsome, P. L. Koch, D. R. Smith, M. E. Church, S. D. Chamberlain, K. J. Sorenson, and R. Risebrough. 2005. Pleistocene to recent dietary shifts in California Condors. Proceedings of the National Academy of Sciences 102:16707–11.

Chamberlain, M. 1887. A Catalogue of Canadian Birds with Notes on the Distribution of the Species. J. & A. McMillian: Saint John, New Brunswick.

Channell, R., and M. V. Lomolino. 2000. Dynamic biogeography and conservation of endangered species. Letters to Nature 403:84–86.

Chapman, D. W. 1986. Salmon and steelhead abundance in the Columbia River in the nineteenth century. Transactions of the American Fisheries Society 115:662–70.

Chase, A. W. 1869. The sea-lion at home. Overland Monthly and Out West Magazine 3:350–54.

Church, M. E., R. Gwiazda, R. W. Risebrough, K. Sorenson, C. P. Chamberlain, S. Farry, W. Heinrich, B. A. Rideout, D. R. Smith. 2006. Ammunition is the principal source of lead accumulated by California Condors re-introduced to the wild. Environmental Science & Technology 40:6143–50.

Clapham, P. J., S. Leatherwood, I. Szczepaniak, R. L. Brownell Jr. 1997. Catches of humpback and other whales from shore stations at Moss Landing and Trinidad, California, 1919–1926. Marine Mammal Science 13:368–94.

Clark, E. E. 2003. Indian Legends of the Pacific Northwest. University of California Press: Berkeley.

Clark, T. D. (editor). 1967. Gold Rush Diary: Being the Journal of Elisha Douglass Perkins on the Overland Trail in the Spring and Summer of 1849. University of Kentucky: Lexington.

Clarke, C., in litt. 1971. March 1971 letter from Cecile Clarke to Sanford Wilbur regarding a possible condor sighting in Mendocino County, CA.

Cleland, R. G. 1951. Cattle on a Thousand Hills: Southern California, 1850–1880, 2nd ed. Huntington Library Press: San Marino, California.

Clyman, J. 1926. James Clyman, his diary and reminiscences. California Historical Quarterly 5:136–87.

Coates, P. 1999. 'Unusually cunning, vicious, and treacherous': the extermination of the wolf in United States history. Pages 163–83 in M. Levene and P. Roberts (editors), The Massacre in History. Berghahn Books: New York.

Cobb, J. N. 1921. Pacific Salmon Fisheries, Appendix 1 to the Report of the US Commissioner of Fisheries for 1921, 3rd ed. Bureau of Fisheries Document 902. Government Printing Office: Washington, DC.

Collier, M. E. T., and S. B. Thalman (editors). 1991. Interviews with Tom Smith and Maria Copa: Isabel Kelly's ethnographic notes on the Coast Miwok Indians of Marin and southern Sonoma Counties, California. Miwok Archeological Preserve of Marin Occasional Papers 6: San Rafael, California.

Cook, S. F. 1956. The aboriginal population of the north coast of California. Anthropological Records 16:81–130.

Cooper, J. G. 1870. Ornithology [of California], vol. 1, Land Birds: Edited by S. F. Baird, From the Manuscript and Notes of J. G. Cooper. Geological Survey of California: Sacramento, California.

Cooper, J. G. 1890. A doomed bird. Zoe 1:248–49.

Cooper, J. G., and G. Suckley. 1859. The Natural History of Washington Territory, with Much Relating to Minnesota, Nebraska, Kansas, Oregon, and California, Between the Thirty-Sixth and Forty-Ninth Parallels of Latitude, Being Those Parts of the Final Reports on the Survey of the Northern Pacific Railroad Route, Containing the Climate and Physical Geography, with Full Catalogues and Descriptions of the Plants and Animals Collected from 1853 to 1857. Bailliére Brothers: New York.

Corbin, G. A. 1988. Native Arts in North America, Africa, and the South Pacific: An Introduction. Harper & Row: New York.

Coues, E. 1897. New Light on the Early History of the Greater Northwest; The Manuscript Journals of Alexander Henry, Fur Trader of the Northwest Company and of David Thompson, Official Geographer and Explorer of the Same Company, 1799–1814. Francis P. Harper: New York.

Craighead, D., and B. Bedrosian. 2008. Blood lead levels of Common Ravens with access to big-game offal. Journal of Wildlife Management 72:240–45.

Crane, H. R. 1956. University of Michigan radiocarbon dates I. Science 124:664–72.

Crane, H. R., and J. B. Griffon. 1960. University of Michigan radiocarbon dates V. American Journal of Science Radiocarbon Supplement 2:31–48.

Cronise, T. F. 1868. The Natural Wealth of California: Comprising Early History; Geography, Topography, and Scenery; Climate; Agriculture and Commercial Products; Geology, Zoology, and Botany; Mineralogy, Mines, and Mining Processes; Manufactures; Steamship Lines, Railroads, and Commerce; Immigration, Population and Society; Educational Institutions and Literature; Together with a Detailed Description of Each County; Its Topography, Scenery, Cities and Towns, Agricultural Advantages, Mineral Resources, and Varied Productions. H. H. Bancroft & Company: San Francisco.

Cummins, E. S. 1893. Story of the Files: A Review of Californian Writers and Literature. World's Fair Commission of California, Columbian Exposition, 1893. Co-operative Printing Co.: San Francisco.

Curtis, E. S. 1910. The North American Indian: Being a Series of Volumes Picturing and Describing the Indians of the United States and Alaska, vol. 7. E. S. Curtis: Seattle; The University Press: Cambridge, Massachusetts.

Curtis, E. S. 1911. The North American Indian: Being a Series of Volumes Picturing and Describing the Indians of the United States and Alaska, vol. 8. E. S. Curtis: Seattle; The University Press: Cambridge, Massachusetts.

Curtis, E. S. 1913. The North American Indian: Being a Series of Volumes Picturing and Describing the Indians of the United States, the Dominion of Canada, and Alaska, vol. 9. E. S. Curtis: Seattle; The University Press: Cambridge, Massachusetts.

Curtis, E. S. 1915. The North American Indian: Being a Series of Volumes Picturing and Describing the Indians of the United States, the Dominion of Canada, and Alaska, vol. 10. E. S. Curtis: Seattle; The University Press: Cambridge, Massachusetts.

Curtis, E. S. 1924. The North American Indian: Being a Series of Volumes Picturing and Describing the Indians of the United States, the Dominion of Canada, and Alaska, vol. 13. E. S. Curtis: Seattle; The University Press: Cambridge, Massachusetts.

Daggett, F. S. 1898. Capture of a California Condor. Osprey 2:134.

Daily Alta California (San Francisco newspaper). 1858. Vulture shot. 4 February.

Daily Evening Bulletin (San Francisco newspaper). 1873. Pacific slope brevities. 19 February.

Daily Evening Bulletin (San Francisco newspaper). 1880. State news in brief. 7 May.

Daily Union (Sacramento newspaper). 1854. California Vulture. 11 March.

Daily Union (Sacramento newspaper). 1854. A California Vulture. 21 June.

Daily Union (Sacramento newspaper). 1856. Nomen Lacken Reservation. 1 April.

Daily Union (Sacramento newspaper). 1857. Decidedly voracious. 24 September.

Daily Union (Sacramento newspaper). 1861. A large bird. 18 June.

Daily Union (Sacramento newspaper). 1865. A huge bird. 25 November.

Daily Union (Sacramento newspaper). 1871. A monster. 26 August.

Davis, L. 1989. Jedediah Smith through Hupa territory. American Indian Quarterly 13:369–89.

Dawson, W. L. 1923. The Birds of California. South Moulton Co.: San Diego, California.

de Lucas, M., M. Ferrer, M. J. Bechard, and A. Muñoz. 2012. Griffon Vulture mortality at wind farms in southern Spain: Distribution of fatalities and active mitigation measures. Biological Conservation 147:184–89.

Díaz, D., M. Cuesta, T. Abreu, and E. Mujjica. 2000. Estrategia de conservación para el Cóndor Andino (*Vultur gryphus*). WWF and Fundación BioAndian: Caracas, Venezuela.

Dillon, R. H. 1961. California Trail Herd: The 1850 Missouri-to-California Journal of Cyrus C. Loveland. Talisman Press: Los Gatos, California.

Dixon, J. 1924. California Condors breed in captivity. Condor 26:192.

Dixon, S. L., and R. L. Lyman. 1996. On the Holocene history of elk (*Cervus elaphus*) in eastern Washington. Northwest Science 70:262–72.

Doherty, K. 2000. Explorers, Missionaries, and Trappers: Trailblazers of the West. Oliver Press: Minneapolis, Minnesota.

Doney, A. E., P. Klink, and W. Russell. 1916. Early game conditions in Siskiyou County. California Fish and Game 2:123–25.

Douglas, D. 1829. Art. 43. Observations on the *Vultur Californianus* of Shaw. Zoological Journal 4:328–30.

Douglas, D. 1914. Journal Kept by David Douglas During His Travels in North America 1823–1827. William Wesley & Son: London.

Drucker, P. 1937. The Tolowa and their southwest Oregon kin. University of California Publications in American Archaeology and Ethnology. 36:221–300.

Du Bois, C. 1935. Wintu ethnography. University of California Publications in American Archaeology and Ethnology 36:1–148.

Ehrlich, C. 1937. Tribal culture in Crow mythology. Journal of American Folklore 50:307–408.

Eisenmann, E. 1963. Is the Black Vulture migratory? Wilson Bulletin 75:244–49.

Emerson, W. O. 1899. Dr. James G. Cooper: A sketch. Condor 1:1–5.

Emslie, S. D. 1987. Age and diet of fossil California Condors in Grand Canyon, Arizona. Science 237:768–70.

Emslie, S. D. 1988. The fossil history and phylogenetic relationships of condors (Ciconiiformes: Vulturidae) in the New World. Journal of Vertebrate Paleontology 8:212–28.

Emslie, S. D. 1990. Additional ^{14}C dates on fossil California Condor. National Geographic Research 6:134–35.

Fannin, J. 1891. Check List of British Columbia Birds. Richard Wolfenden: Victoria, British Columbia.

Fannin, J. 1897. The California Vulture in Alberta. Auk 14:89.

Feranec, R. S. 2009. Implications of radiocarbon dates from Potter Creek Cave, Shasta County, California, USA. Radiocarbon 51:931–36.

Feranec, R. S., E. A. Hadly, J. L. Blois, and A. D. Barnosky. 2007. Radiocarbon dates from the Pleistocene fossil deposits of Samwel Cave, Shasta County, California, USA. Radiocarbon 49:117–21.

Ferguson-Lees, J. and D. A. Christie. 2001. Raptors of the World. Houghton Mifflin: New York.

Ficken, R. E. 1987. The Forested Land: A History of Lumbering in Western Washington. University of Washington Press: Seattle.

Finkelstein, M. E., D. F. Doak, D. George, J. Burnett, J. Brandt, M. Church, J. Grantham, and D. R. Smith. 2012. Lead poisoning and the deceptive recovery of the critically endangered California Condor. Proceedings of the National Academy of Sciences, DOI: 10.1073/pnas.1203141109.

Finkelstein, M. E., D. George, S. Scherbinski, R. Gwiazda, M. Johnson, J. Burnett, J. Brandt, S. Lawrey, A. P. Pessier, M. Clark, J. Wynne, J. Grantham, and D. R. Smith. 2010. Feather lead concentrations and ^{207}Pb/^{206}Pb ratios reveal lead exposure history of California Condors (*Gymnogyps californianus*). Environmental Science & Technology 44:2639–47.

Finley, E. L. (editor). 1937. History of Sonoma County, California: Its People and Its Resources. Press Democrat Publishing Company: Santa Rosa, California.

Finley, W. L. 1908a. Home life of the California Condor: A record of unique personal experience while making at close range the first photographs of the bird. The Century Magazine 75:370–80.

Finley, W. L. 1908b. Life history of the California Condor, Part 2.—Historical data and range of the condor. Condor 10:5–10.

Finley, W. L. 1910. Life history of the California Condor, Part 4. —The young condor in captivity. Condor 12:4–11.

Finley, W. L. 1941. California Vulture. National Association of Audubon Societies' Educational Leaflet 123:489–92.

Finley, W. L., and I. Finley. 1915. The condor as a pet. Bird-Lore 17:413–19.

Fisher, A. K. 1920. In memoriam: Lyman Belding. Auk 37:33–45.

Fisher, H. I. 1946. Adaptations and comparative anatomy of the locomotor apparatus of New World vultures. American Midland Naturalist 35:545–727.

Fleming, J. H. 1924. The California Condor in Washington: Another version of an old record. Condor 26:111–12.

Forest and Stream. 1895. First annual sportsman's exposition, Madison Square Garden, New York, May 13 to 18. Forest and Stream 44:417–30.

Fox-Dobbs, K., T. A. Stidham, G. J. Bown, S. D. Emslie, and P. L. Koch. 2006. Dietary controls on extinction versus survival among avian megafauna in the late Pleistocene. Geology 34:685–88.

Fry, D. M. 2003. Assessment of Lead Contamination Sources Exposing California Condors. Final Report. Submitted to California Department of Fish and Game, Sacramento, California, April 7, 2003.

Fry, W. 1928. The California Condor: A modern Roc, the Arabian Nights come true in California. Sequoia National Park Bulletin 23, February 10, 1928.

Fuller, E. 2001. Extinct Birds, revised ed. Comstock Books: Ithaca, New York.

Gabrielson, I. N., and S. G. Jewett. 1970. Birds of the Pacific Northwest: With Special Reference to Oregon (Originally titled Birds of Oregon and printed by Oregon State College in 1940). Dover Publications: New York.

Gambel, W. 1847. Remarks on the birds observed in Upper California, with descriptions of new species. Proceedings of the Natural Academy of Sciences of Philadelphia 1:25–56.

García-Ripollés, C., P. López-López, and V. Urios. 2010. First description of migration and wintering of adult Egyptian vultures *Neophron percnopterus* tracked by GPS satellite telemetry. Bird Study 57:261–65.

Garth, T. R. 1953. Atsugewi ethnography. Anthropological Records 14:129–212.

Gassaway, F. H. 1882. Summer Saunterings, by "Derrick Dodd." Francis, Valentine & Company: San Francisco.

Gaston, J. 1912. The Centennial History of Oregon, 1811–1912, vol. 1. S. J. Clarke Publishing Company: Chicago.

Gatschet, A. S. 1890. Dictionary of the Klamath language. Contributions to North American Ethnology, vol. 2. Government Printing Office: Washington, DC.

Gerber, L. R., D. P. DeMaster, and S. P. Roberts. 2000. Measuring success in conservation: Assessing efforts to restore populations of marine mammals is partly a matter of epistemology: How do you know when enough is enough? American Scientist 88:316–24.

Gerlach, A. C. (editor). 1970. The National Atlas of the United States of America. US Geological Survey, Department of the Interior: Washington, DC.

Gibson, J. R. 1985. Farming the Frontier: The Agricultural Opening of the Oregon Country, 1786–1846. University of Washington Press: Seattle.

Gifford, E. W. 1940. California bone artifacts. Anthropological Records 3:153–237.

Gifford, E. W. 1955. Central Miwok ceremonies. Anthropological Records 14:261–318.

Gilbert, M., M. Z. Virani, R. T. Watson, J. L. Oaks, P. C. Benson, A. A. Khan, S. Ahmed, J. Chaudhry, M. Arshad, S. Mahmood, and Q. A. Shah. 2002. Breeding and mortality of Oriental White-backed Vulture *Gyps bengalensis* in Punjab Province, Pakistan. Bird Conservation International 12:311–26.

Grayson, D. K. 2011. The Great Basin: A Natural Prehistory. University of California Press: Berkeley.

Green, R. E., I. Newton, S. Shultz, A. A. Cunningham, M. Gilbert, D. J. Pain, and V. Prakash. 2004. Diclofenac poisoning as a cause of vulture population declines across the Indian subcontinent. Journal of Applied Ecology 41:793–800.

Griffin, M. C., K. Entriken, J. Hodel, and T. C. Weston. 2006. Osteological analysis of the human skeletal remains from the Filoli Site, San Mateo County, California (CA-SMA-125). Society for California Archaeology Newsletter 40:32–35.

Griffith, B., J. M. Scott, J. W. Carpenter, C. Reed. 1989. Translocation as a species conservation tool: Status and strategy. Science 245:477–79.

Grinnell, J. 1913. The status of the California Condor in 1912. Pages 22–24 *in* Our Vanishing Wild Life: Its Extermination and Preservation, by W. T. Hornaday. Charles Scribner's Sons: New York.

Hackett, S. J., R. T. Kimball, S. Reddy, R. C. K. Bowie, E. L. Braun, M. J. Braun, J. L. Chojnowski, W. A. Cox, K.-L. Han, J. Harshman, C. J. Huddleston, B. D. Marks, K. J. Miglia, W. S. Moore, F. H. Sheldon, D. W. Steadman, C. C. Witt, and T. Yuri. 2008. A phylogenomic study of birds reveals their evolutionary history. Science 320:1763–68.

Hall, F. S. 1934. Studies of the history of ornithology in the State of Washington (1792–1932) with special reference to the discovery of new species; Part 3: David Douglas, pioneer naturalist on the Columbia River—1825–1833. Murrelet 15:2–19.

Hammond, L. 2006. Marketing wildlife: The Hudson's Bay Company and the Pacific Northwest, 1821–1849. Pages 203–22 *in* D. F. Duke (editor), Canadian Environmental History: Essential Readings. Canadian Scholars' Press: Toronto, Ontario.

Hampton, B. 1997. The Great American Wolf. Henry Holt and Company: New York.

Hansel-Kuehn, V. J. 2003. The Dalles Roadcut (Fivemile Rapids) avifauna: Evidence for a cultural origin. M.S. Thesis, Washington State University, Department of Anthropology. May 2003.

Hansen, H. P. 1947. Postglacial forest succession, climate, and chronology in the Pacific Northwest. Transactions of the American Philosophical Society 37:1–130.

Harper, J. A., J. H. Harn, W. W. Bently, and C. F. Yocom. 1967. The status and ecology of the Roosevelt elk in California. Wildlife Monographs 16:1–49.

Harpole, J. L., and R. L. Lyman. 1999. The Holocene biogeographic history of elk (*Cervus elaphus*) in western Washington. Northwest Science 73:106–13.

Harrington, J. P. 1932. Karuk Indian Myths. Smithsonian Institution Bureau of American Ethnology Bulletin 107. US Government Printing Office: Washington, DC.

Harris, H. 1941. The annals of *Gymnogyps* to 1900. Condor 43:3–55.

Harrison, E. N., and L. F. Kiff. 1980. Apparent replacement clutch laid by wild California Condor. Condor 82:351–52.

Helander, B, J. Axelsson, H. Borg, K. Holm, and A. Bignert. 2009. Ingestion of lead from ammunition and lead concentrations in White-tailed Sea Eagles (*Haliaeetus albicilla*) in Sweden. Science in the Total Environment 407:5555–63.

Hendrickson, S. L., R. Bleiweiss, J. C. Matheus, L. S. de Matheus, N. L. Jácome, and E. Pavez. 2003. Low genetic variability in the geographically widespread Andean Condor. Condor 105:1–12.

Henshaw, H. W. 1876. Appendix H8: Report on the Ornithology of the Portions of California Visited During the Field-season of 1875. Pages 444–520 *in* Report of the Secretary of War; Being Part of the Message and Documents Communicated to the Two Houses of Congress at the Beginning of the Second Session of the Forty-fourth Congress, vol. 2, part 3. Government Printing Office: Washington, DC.

Henshaw, T. 1993. The History of Winchester Firearms, 1866–1992. 6th ed. Winchester Press: Clinton, New Jersey.

Hernández, M., and A. Margalida. 2008. Pesticide abuse in Europe: Effects on the Cinereous Vulture (*Aegypius monachus*) population in Spain. Ecotoxicology 17:264–72.

Hernández, M., and A. Margalida. 2009. Poison-related mortality effects in the endangered vulture (*Neophron percnopterus*) population in Spain. European Journal of Wildlife Research, DOI 10.1007/s10344-009-0255-6.

Hertel, F. 1995. Ecomorphological indicators of feeding behavior in recent and fossil raptors. Auk 112:890–903.

Hessburg, P. F., and J. K. Agee. 2003. An environmental narrative of Inland Northwest forests, 1800–2000. Forest Ecology and Management 178:23–59.

Hildebrandt, W. R., and T. L. Jones. 1992. Evolution of marine mammal hunting: A view from the California and Oregon coasts. Journal of Anthropological Archaeology 11:360–401.

Hines, H. K. 1894. An Illustrated History of the State of Washington. Lewis Publishing Company: Chicago.

Hodgson, S. F. 2007. Obsidian: Sacred glass from the California sky. Pages 295–313 in L. Piccardi and W. B. Masse (editors), Myth and Geology. Geological Society, London, Special Publications 273.

Holmes, F. H. 1897. A pet condor. Nidiologist 4:58–59.

Holyoake, M. Q. 1891. Captain Mayne Reid: Soldier and novelist. The Strand Magazine 2:93–102.

Houston, D. C. 1975. Ecological isolation of African scavenging birds. Ardea 63:55–64.

Houston, D. C. 1979. The adaptations of scavengers. Pages 263–86 in A. R. E. Sinclair and M. N. Griffiths (editors), Serengeti: Dynamics of an Ecosystem. University of Chicago Press: Chicago.

Houston, D. C. 1983. The adaptive radiation of the Griffon Vultures. Pages 135–52 in S. R. Wilbur and J. A. Jackson (editors), Vulture Biology and Management. University of California Press: Berkeley.

Houston, D. C. 1985. Evolutionary ecology of Afrotropical and Neotropical vultures in forests. Ornithological Monographs 36:856–64.

Houston, D. C. 1988. Competition for food between Neotropical vultures in forest. Ibis 130:402–17.

Howard, H. 1929. The avifauna of Emeryville Shellmound. University of California Publications in Zoology 32:301–94.

Howard, H. 1952. The prehistoric avifauna of Smith Creek Cave, Nevada, with a description of a new gigantic raptor. Bulletin of the Southern California Academy of Sciences 51:50–54.

Hunn, E. S. 1990. Nch'i-Wána, "The Big River": Mid-Columbia Indians and Their Land. University of Washington Press: Seattle.

Hunt, W. G., C. N. Parish, S. C. Farry, T. G. Lord, and R. Sieg. 2007. Movements of introduced California Condors in Arizona in relation to lead exposure. Pages 79–96 in A. Mee and L. S. Hall (editors), California Condors in the 21st Century. Series in Ornithology 2. Nuttall Ornithological Club and American Ornithologists' Union: Cambridge, Massachusetts, and Washington, DC.

Hunt, W. G., R. T. Watson, J. L. Oaks, C. N. Parish, K. K. Burnham, R. L. Tucker, J. R. Belthoff, and G. Hart. 2009. Lead bullet fragments in venison from rifle-killed deer: Potential for human dietary exposure. PloS ONE 4(4): e5330, DOI:10.1371/journal.pone.0005330.

I. R. Wilson Consultants, Ltd. 2006. Poets Cove, Bedwell Harbour, Pender Island, B.C., Heritage Site Alteration Permit 2002-388. Report on file at the BC Archaeology Branch: Victoria, British Columbia.

IUCN (International Union for Conservation of Nature). 1987. IUCN guidelines for re-introductions. Prepared by the IUCN/SSC Re-introduction Specialists Group. IUCN: Gland, Switzerland and Cambridge, United Kingdom.

Jackson, J. A., I. D. Prather, R. N. Conner, and S. P. Gaby. 1978. Fishing behavior of Black and Turkey Vultures. The Wilson Bulletin 90:141–43.

Janssen, D. L., and M. P. Anderson. 1986. Lead poisoning in free-ranging California Condors. Journal of the American Veterinary Association 189:1115–17.

Jewett, S. G., W. P. Taylor, W. T. Shaw, and J. W. Aldrich. 1953. Birds of Washington State. University of Washington Press: Seattle.

Jobanek, G. A., and D. B. Marshall. 1992. John K. Townsend's 1836 report of the birds of the lower Columbia River region, Oregon and Washington. Northwestern Naturalist 73:1–14.

Johnson, E. V., D. L. Aulman, D. A. Clendenen, G. Guliasi, L. M. Morton, P. I. Principe, and G. M. Wegener. 1983. California Condor: Activity patterns and age composition in a foraging area. American Birds 37:941–45.

Johnson, J. J. 1967. The archaeology of the Camanche Reservoir locality, California. Sacramento Anthropological Society Papers 6:1–346.

Johnson, M., J. Kern, and S. M. Haig. 2010. Analysis of California Condor (*Gymnogyps californianus*) use of six management units using location data from global positioning system transmitters, southern California, 2004–2009—Initial Report. US Geological Survey Open-File Report 2010-1287.

Johnson, M. L., and S. Johnson. 1952. Check list of mammals of the Olympic Peninsula. The Murrelet 33:32–37.

Jollie, M. 1953. The birds observed in Idaho by the Lewis and Clark expedition, 1804–1806. The Murrelet 34:1–5.

Jones, K. 2002. Wolf Mountains: A History of Wolves along the Great Divide. University of Calgary Press: Calgary, Alberta.

Kay, C. E. 2007. Were native people keystone predators? A continuous-time analysis of wildlife observations made by Lewis and Clark in 1804–1806. The Canadian Field-Naturalist 121:1–15.

Kehoe, T., and C. Jacobson. 2003. Environmental decision making and DDT production at Montrose Chemical Corporation of California. Enterprise and Society 4:640–75.

Kenyon, K. W., and V. B. Scheffer. 1961. Wildlife surveys along the northwest coast of Washington. 1961. The Murrelet 42:29–37.

Kidd, T. 1927. The History of Lulu Island and Occasional Poems. Reprinted by the City of Richmond Archives (2007): Richmond, British Columbia.

Kiff, L. F. 1989. DDE and the California Condor *Gymnogyps californianus*: The end of a story? Pages 477–80 *in* B.-U. Myburg and R. D. Chancellor (editors), Raptors in the Modern World. WWGBP: Berlin, London, and Paris.

Kiff, L. F. 2005. History, present status, and future prospects of avian eggshell collections in North America. Auk 122:994–99.

Kiff, L. F., D. B. Peakall, and S. R. Wilbur. 1979. Recent changes in California Condor eggshells. Condor 81:166–72.

Kim, J.-H., O.-S. Chung, W.-S. Lee, and Y. Kanai. 2007. Migration routes of Cinereous Vultures (*Aegypius monachus*) in northeast Asia. Journal of Raptor Research 41:161–65.

Kingston, C. S. 2010. Buffalo in the Pacific Northwest. Washington Historical Quarterly 23:163–72.

Kirk, D. A. and M. J. Mossman. 1998. Turkey Vulture (*Cathartes aura*), The Birds of North America Online (A. Poole, editor). Cornell Lab of Ornithology: Ithaca, New York, http://bna.birds.cornell.edu/bna/species/339.

Kneubuehl, B. P., R. M. Coupland, M. A. Rothschild, and M. J. Thali. 2006. Wound Ballistics: Basics and Applications. Springer-Verlag: Berlin.

Koch, G. 2004. California Condors return to Oregon. Association of Zoo and Aquarium Docents (AZAD) 2004 Conference Paper, Philadelphia Zoo, Philadelphia, PA. Available online at http://azadocents.org/wordpress/?p=198.

Koford, C. B. 1941. Unpublished notes, April 11, 1941. Museum of Vertebrate Zoology, Berkeley, California.

Koford, C. B. 1953. The California Condor. Research Report 4. National Audubon Society: New York.

Kroeber, A. L. 1908. Wiyot folk-lore. Journal of American Folklore 21:37–39.

Kroeber, A. L. 1925. Handbook of the Indians of California, vol. 1. Bulletin of the Bureau of American Ethnology of the Smithsonian Institution. Smithsonian Institution: Washington, DC.

Kroeber, A. L. 1929. The Valley Nisenan. University of California Publications in American Archaeology and Ethnology 24:253–90.

Kroeber, A. L. 1976. Yurok Myths. University of California Press: Berkeley.

Kroeber, A. L., and E. W. Gifford. 1949. World renewal: A cult system of native northwest California. Anthropological Records 13:1–156.

Kroeber, A. L., and E. W. Gifford. 1980. Karok Myths. University of California Press: Berkeley.

Laliberte, A. S., and W. J. Ripple 2004. Range contractions of North American carnivores and ungulates. BioScience 54:123–38.

Lambertucci, S. A. 2007. Biología y conservación del Cóndor Andino (*Vultur gryphus*) en Argentina. Hornero 22:149–58.

Lambertucci, S. A. 2010. Size and spatio-temporal variations of the Andean condor *Vultur gryphus* population in north-west Patagonia, Argentina: Communal roosts and conservation. Oryx 44:441–47.

Lambertucci, S. A., J. A. Donazar, A. D. Huertas, B. Jiminez, M. Saez, J. A. Sanchez-Zapata, and F. Hiraldo. 2011. Widening the problem of lead poisoning to a South-American top scavenger: Lead concentrations in feathers of wild Andean Condors. Biological Conservation 144:1464–71.

Lamson, J. 1852–1861. Nine Years' Adventure in California, From September 1852 to September 1861, with Excursions into Oregon, Washington, and Nevada. Manuscript Diary, California Historical Society, San Francisco.

Lamson, J. 1878. Round Cape Horn. Voyage of the Passenger-ship James W. Paige, from Maine to California in the Year 1852. O. F. & W. H. Knowles: Bangor, Maine.

Leach, F. A. 1929. A Turkey Buzzard roost. Condor 31:21–23.

Lewis, M., W. Clark, and members of the Corps of Discovery. 2002. The Journals of the Lewis and Clark Expedition (G. Moulton, editor). University of Nebraska Press: Lincoln. University of Nebraska-Lincoln Libraries Electronic Text Center, The Journals of the Lewis and Clark Expedition, http://lewisandclarkjournals.unl.edu/.

Libby, W. F. 1954. Chicago radiocarbon dates V. Science 120:733–42.

Lichatowich, J., L. Mobrand, and L. Lestelle. 1999. Depletion and extinction of Pacific salmon (Oncorhynchus spp.): A different perspective. ICES Journal of Marine Science 56: 467–72.

Lieberson, P. 1988. Appendix 4: Faunal analysis of CA-SFR-112. Pages 117–29 in: A. G. Pastron and M. R. Walsh (editors), Archaeological Excavations at CA-SFR-112, the Stevenson Street Shellmound, San Francisco, California. Coyote Press Archives of California Prehistory 21: Salinas, California.

Linsdale, J. M. 1931. Facts concerning the use of thallium in California to poison rodents: Its destructiveness to game birds, song birds and other valuable wild life. Condor 33:92–106.

Linsdale, J. M. 1932. Facts concerning losses to wild animal life through pest control in California. Condor 34:121–35.

Lipchitz, B. A. 1955. On the collections of the Museum of Anthropology and Ethnography gathered by Russian travelers and explorers in Alaska and California [in Russian]. Collections of the Museum of Anthropology and Ethnography 16:358–69.

Livezey, B. C., R. L. Zusi. 2007. Higher-order phylogeny of modern birds (Theropoda, Aves: Neornithes) based on comparative anatomy. II. Analysis and discussion. Zoological Journal of the Linnean Society 149:1–95.

Locke, L. N., G. E. Bagley, D. N. Frickie, and L. T. Young. 1969. Lead poisoning and aspergillosis in an Andean Condor. Journal of the American Veterinary Medical Association 155:1052–56.

Loeb, E. M. 1926. Pomo Folkways. University of California Publications in American Archaeology and Ethnology 19:149–405.

Loeb, E. M. 1932. The western Kuksu cult. University of California Publications in American Archaeology and Ethnology 33:1–137.

Loeb, E. M. 1933. The eastern Kuksu cult. University of California Publications in American Archaeology and Ethnology 33:139–232.

Longhurst, W. M., A. S. Leopold, and R. F. Dasmann. 1952. A survey of California deer herds, their ranges and management problems. California Department of Fish and Game, Game Bulletin 6: Sacramento, California.

Lord, J. K. 1866. The Naturalist in Vancouver Island and British Columbia, in Two Volumes—vol. 2. Richard Bentley: London.

Love, C. M. 1916. History of the cattle industry in the Southwest. The Southwest Historical Quarterly 19:370–99.

Lowie, R. H. 1918. Myths and traditions of the Crow Indians. Anthropological Papers of the American Museum of Natural History, vol. 25, part 1. New York.

Lucas, F. A. 1891. Animals recently extinct or threatened with extermination, as represented in the collections of the U.S. National Museum. Pages 609–49 in Report of the National Museum: Annual Report of the Board of Regents of the Smithsonian Institution Showing the Operations, Expenditures, and Condition of the Institution For the Year Ending June 30, 1889. Government Printing Office: Washington, DC.

Lyman, R. E. 2006. Late prehistoric and early historic abundance of Columbian white-tailed deer, Portland Basin, Washington and Oregon, USA. Journal of Wildlife Management 70:278–82.

Lyman, R. L., J. L. Harpole, C. Darwent, and R. Church. 2002. Prehistoric occurrence of pinnipeds in the lower Columbia River. Northwest Naturalist 83:1–6.

Lyman, R. L., and S. Wolverton. 2002. The late prehistoric-early historic game sink in the northwestern United States. Conservation Biology 16:73–85.

Lyon, M. W. 1918. Biological Society of Washington, Report of the Secretary, October 20, 1917. Journal of the Washington Academy of Sciences 8:25–28.

Mace, M. 2011. California Condor International Studbook (Gymnogyps californianus), January 31, 2011. Compiled by Michael Mace, Zoological Society of San Diego.

Macoun, J. 1903. Catalogue of Canadian Birds, Part 2: Birds of Prey, Woodpeckers, Fly-catchers, Crows, Jays, and Blackbirds. Geological Survey of Canada: Ottawa.

Maloney, A. B. 1945. Fur brigade to Bonaventura; John Work's California expedition, 1832–1833, for the Hudson's Bay Company. California Historical Society: San Francisco.

Mandel, J. T., K. L. Bildstein, G. Bohrer, and D. W. Winkler. 2008. Movement ecology of migration in Turkey Vultures. Proceedings of the National Academy of Sciences of the United States of America 105:19102–7.

Maniery, J. G. 1991. Mokelumne River and Tributaries, California: Cultural Resources Summary. Prepared by Par Environmental Services, Inc., for the US Army Corps of Engineers, Sacramento District, Planning Division, Contract No. DACW0590P2692. Sacramento, California.

Mark, S. R. 2006. Domain of the Caveman: A Historic Resource Study of Oregon Caves National Monument. National Park Service, Pacific West Region: Oakland, California.

Marshall, W. I. 1911. Acquisition of Oregon and the Long Suppressed Evidence about Marcus Whitman, Part 2. Lowman & Hanford Co.: Seattle.

Martin, P. S. 1990. 40,000 years of extinctions on the "planet of doom." Palaeogeography, Palaeoclimatology, Palaeoecology 82:187–201.

Martin, P. S., and C. R. Szuter. 1999. Megafauna of the Columbia Basin, 1800–1840: Lewis and Clark in a game sink. Pages 188–204 in D. D. Goble and P. W. Hirt (editors), Northwest Lands, Northwest Peoples: Readings in Environmental History. University of Washington Press: Seattle.

Matteson, F. S. 1886. Insects and their enemies. *Willamette Farmer* (Salem, Oregon) 25 June.

Matthewson, W. 1986. William L. Finley, Pioneer Wildlife Photographer. Oregon State University Press: Corvallis.

Mattina, A. 1987. Colville-Okanagan Dictionary. University of Montana Occasional Papers in Linguistics 5: Missoula.

Maunder, W. J. 1968. Synoptic weather patterns in the Pacific Northwest. Northwest Science 42:80–88.

McCornack, E. C. 1920. Contributions to the Pleistocene history of Oregon. University of Oregon Leaflet Series, Geology Bulletin 6:3–23.

McCullough, D. R. 1969. The tule elk: Its history, behavior, and ecology. University of California Publications in Zoology 88:1–209.

McGahan, J. 1973. Flapping flight of the Andean Condor in nature. Journal of Experimental Biology 58:239–53.

McGahan, J. 2012. The Andean Condor: A Field Study. Self-published: Arlee, Montana. Available online at the Smithsonian Institution Archives: http://archive.org/details/TheAndeanCondorAFieldStudy.

McKelvey, K. S., K. B. Aubry, and M. K. Schwartz. 2008. Using anecdotal occurrence data for rare or elusive species: The illusion of reality and a call for evidentiary standards. BioScience 58:549–55.

McKelvey, K. S., and J. D. Johnson. 1992. Historical perspectives on forests of the Sierra Nevada and the Transverse Ranges of southern California: Forest conditions at the turn of the century. Pages 225–46 (chap. 11) in The California Spotted Owl: A Technical Assessment of its Current Status. USDA Forest Service, Pacific Southwest Research Station General Technical Report PSW-GTR-133. Albany, California.

McKernan, D. L., D. R. Johnson, and J. I. Hodges. 1950. Some factors influencing the trends of salmon populations in Oregon. Oregon Fish Commission Contribution 12:427–49. Portland, Oregon. Reprinted from Transactions of the Fifteenth North American Wildlife Conference, March 6, 7, and 9, 1950. Published by the Wildlife Management Institute: Washington, DC.

McMillan, I. 1968. Man and the California Condor: The Embattled History and Uncertain Future of North America's Largest Free-living Bird. E. P. Dutton & Co.: New York.

McRoskey, R. 1914. The Missions of California: With Sketches of the Lives of St. Francis and Junipero Serra. Philopolis Press: San Francisco.

Meany, E. S. 1907. Last survivor of the Oregon mission of 1840. Washington Historical Quarterly 2:12–23.

Mee, A., B. A. Rideout, J. A. Hamber, J. N. Todd, G. Austin, M. Clark, and M. P. Wallace. 2007. Junk ingestion and nestling mortality in a reintroduced population of California Condors *Gymnogyps californianus*. Bird Conservation International 17:119–30.

Meengs, C. C., and R. T. Lackey. 2005. Estimating the size of historical Oregon salmon runs. Reviews in Fisheries Science 13:51–66.

Meinig, D. W. 1998. The Shaping of America: A Geographical Perspective on 500 Years of History. Vol. 3: Transcontinental America, 1850–1915. Yale University Press: New Haven, Connecticut.

Meiri, S., and Y. Yom-Tov. 2004. Ontogeny of large birds: Migrants do it faster. Condor 106:540–48.

Mendelsohn, J. M., A. C. Kemp, H. C. Biggs, R. Biggs, and C. J. Brown. 1989. Wing areas, wing loadings and wing spans of 66 species of African raptors. Ostrich 60:35–42.

Mercer, B. 2005. People of the River: Native Arts of the Oregon Territory. University of Washington Press: Seattle.

Meretsky, V. J., and N. F. R. Snyder. 1992. Range use and movements of California Condors. Condor 94:313–35.

Meretsky, V. J., N. F. R. Snyder, S. R. Beissinger, D. A. Clendenen, and J. W. Wiley. 2000. Demography of the California Condor: Implications for reestablishment. Conservation Biology 14:957–67.

Merriam, C. H. 1897. *Cervus roosevelti*, a new elk from the Olympics. Proceedings of the Biological Society of Washington 11:271–75.

Merriam, C. H. 1908. Totemism in California. American Anthropologist 10:558–62.

Merriam, C. H. (editor). 1910. The Dawn of the World: Myths and Weird Tales Told by the Mewan Indians of California. Arthur H. Clark Company: Cleveland, Ohio.

Meyburg, B.-U., M. Gallardo, C. Meyburg, and E. Dimitrova. 2004. Migrations and sojourn in Africa of Egyptian Vultures (*Neophron percnopterus*) tracked by satellite. Journal of Ornithology 145:273–80.

Miller, A. H. 1942. A California Condor bone from the coast of southern Oregon. The Murrelet 23:77.

Miller, A. H., I. McMillan, and E. McMillan. 1965. The current status and welfare of the California Condor. National Audubon Society Research Report 6. 61 pp.

Miller, J. 1999. Chehalis area traditions, a summary of Thelma Anderson's 1927 ethnographic notes. Northwest Anthropological Research Notes 33:1–72.

Miller, L. 1909. *Teratornis*, a new avian genus from Rancho La Brea. University of California Publications, Bulletin of the Department of Geology 5:305–17.

Miller, L. 1942. Succession in the Cathartine dynasty. Condor 44:212–13.

Miller, L. 1953. More trouble for the California Condor. Condor 55:47–48.

Miller, L. 1957. Bird remains from an Oregon Indian midden. Condor 59:59–63.

Miller, L. H. 1910a. The Condor-like Vultures of Rancho La Brea. University of California Publications, Bulletin of the Department of Geology 6:1–19.

Miller, L. H. 1910b. Fossil birds from the Quaternary of southern California. Condor 12:12–15.

Miller, L. H. 1911. Avifauna of the Pleistocene Cave Deposits of California. University of California Publications, Bulletin of the Department of Geology 6:385–400.

Miller, R. C., E. D. Lumley, and F. S. Hall. 1935. Birds of the San Juan Islands, Washington. The Murrelet 16:51–65.

Millikan, C. 1900. Capture of a condor in El Dorado Co. Cal. in 1854. Condor 2:12–13.

Mineau, P., M. R. Fletcher, L. C. Glaser, N. J. Thomas, C. Brassard, L. K. Wilson, J. E. Elliott, L. A. Lyon, C. J. Henny, T. Bollinger, S. L. Porter. 1999. Poisoning of raptors with organophosphorus and carbamate pesticides with emphasis on Canada, U.S. and U.K. Journal of Raptor Research 33:1–37.

Minto, J. 1905. Wild animals in Oregon. Oregon Teachers' Monthly 10:193–95.

Minto, J. 1908. From youth to age as an American, Chapter 2: Learning to live on the land. Oregon Historical Quarterly 9:127–72.

Moen, D. B. 2008. Condors in the Oregon Country: Examining the past to prepare for the future. M.S. Thesis, Portland State University, Portland, Oregon.

Monks, G. G. 2001. Quit blubbering: An examination of Nuu'chah'nulth (Nootkan) whale butchery. International Journal of Osteoarchaeology 11:136–49.

Monteagle, F. J. 1976. A Yankee Trader in the California Redwoods. East Bay Regional Park District: Oakland, California.

Morejohn, G. V., and J. P. Galloway. 1983. Identification of avian and mammalian species used in the manufacture of bone whistles recovered from a San Francisco Bay area archaeological site. Journal of California and Great Basin Anthropology 5:87–97.

Moss, M. L., and J. M. Erlandson. 2008. Native American archaeological sites of the Oregon coast: The historic context for the nomination to the National Register of Historic Places. Pages 1–36 in G. L. Tasa and B. L. O'Neill (editors), Dunes, Headlands, Estuaries, and Rivers: Current Archaeological Research on the Oregon Coast. Association of Oregon Archaeologists Occasional Papers 8: Eugene, Oregon.

Mundy, P., D. Butchart, J. Ledger, and S. Piper. 1992. The Vultures of Africa. Academic Press: London.

Murphy, J. M. 1879. Rambles in North-western America from the Pacific Ocean to the Rocky Mountains. Chapman and Hall: London.

National Marine Fisheries Service. 1991. Recovery plan for the humpback whale (*Megaptera novaeangliae*). Humpback Whale Recovery Team for the National Marine Fisheries Service: Silver Spring, Maryland.

Newberry, J. S. 1857. Report Upon the Zoology of the Route, vol. 6, part 4 of Explorations and Surveys for a Railroad Route from the Mississippi River to the Pacific Ocean: Routes in California and Oregon Explored by Lieut. R. S. Williamson, Corps of Topographical Engineers, and Lieut. Henry L. Abbot, Corps of Topographical Engineers, in 1855. Government Printing Office: Washington, DC.

New York Times. 1874. Ravages of a pestilence. How the Indians were destroyed in California in 1833. Reprint of a communication from Mr. J. J. Warner in the Los Angeles *Star.* 22 October.

New York Times. 1906. A scientist's 12,000-mile hunt for big game: Prof. Chapman's interesting experiences in his summer search for specimens in the West and Southwest. 26 August.

New York Zoological Society. 1907. An almost extinct bird. Zoological Society Bulletin 24:318–20.

Nicolay, C. G. 1846. Oregon Territory: A Geographical and Physical Account of that Country and its Inhabitants with Outlines of its History and Discovery. Charles Knight & Co.: London.

Nomland, G. A. 1938. Bear River ethnography. Anthropological Records 2:91–124.

Northcote, T. G., and D. Y. Atagi. 1997. Pacific salmon abundance trends in the Fraser River watershed compared with other British Columbia systems. Pages 199–219 *in* D. J. Stouder, P. A. Bisson, and R. J. Naiman (editors), Pacific Salmon and Their Ecosystems: Status and Future Options. Chapman & Hall: New York.

Nuttall, T. 1840. A Manual of the Ornithology of the United States and of Canada. 2nd ed., with additions. Hilliard, Gray, and Company: Boston, Massachusetts.

Oaks, J. L., M. Gilbert, M. Z. Virani, R. T. Watson, C. U. Meteyer, B. A. Rideout, H. L. Shivaprasad, S. Ahmed, M. J. I. Chaudhry, M. Arshad, S. Mahmood, A. Ali, and A. A. Khan. 2004. Diclofenac residues as the cause of vulture population decline in Pakistan. Nature 427:630–33.

Occidens, S. 1886. Indian myths about thunder. Knowledge 10:32–34.

Ogada, D. L., and F. Keesing. 2010. Decline of raptors over a three-year period in Laikipia, central Kenya. Journal of Raptor Research 44:129–35.

Okladnikova, E. A. 1983. The California collection of I. G. Voznesensky and the problems of ancient cultural connections between Asia and America. Journal of California and Great Basin Anthropology 5:224–39.

Oregon Fish and Game Commissioner's Board. 1901. Annual Reports of the Department of Fisheries of the State of Oregon to the Legislative Assembly, Twenty-first Regular Season. W. H. Leeds: Salem, Oregon.

Otieno, P. O., J. O. Lalah, M. Virani, I. O. Jondiko, and K.-W. Schramm. 2010. Carbofuran and its toxic metabolites provide forensic evidence for Furadan exposure in vultures (*Gyps africanus*) in Kenya. Bulletin of Environmental Contamination and Toxicology 84:536–44.

Ott, G. 1971. Thunder in their wings. Northwest Magazine (supplement to the Oregonian), March 28: 6–7.

Parmalee, P. W. 1969. California Condor and other birds from Stanton Cave, Arizona. Journal of the Arizona Academy of Science 5:204–6.

Paterek, J. 1996. Encyclopedia of American Indian Costume. W. W. Norton & Company, New York.

Paterson, R. L., Jr. 1984. High incidence of plant material and small mammals in the autumn diet of Turkey Vultures in Virginia. Wilson Bulletin 96:467–69.

Pattee, O. H., P. H. Bloom, J. M. Scott, and M. R. Smith. 1990. Lead hazards within the range of the California Condor. Condor 92:931–37.

Paulbitski, P. A. 1974. Pinnipeds observed in rivers of northern California. California Fish and Game 60:48–49.

Pauli, J. N., and S. W. Buskirk. 2007. Recreational shooting of prairie dogs: A portal for lead entering wildlife food chains. Journal of Wildlife Management 71:102–8.

Pauly, D. 1995. Anecdotes and the shifting baseline syndrome of fisheries. Trends in Ecology and Evolution 10:430.

Payen, L. A., and R. E. Taylor. 1976. Man and the Pleistocene fauna at Potter Creek Cave, California. Journal of California Anthropology 3:51–58.

Peale, T. R. 1848. Mammalia and Ornithology. Vol. 8, United States Exploring Expedition during the years 1838, 1839, 1840, 1841, 1842 under the command of Charles Wilkes, U. S. N. C. Sherman: Philadelphia.

Peck, G. D. 1904. The Cal. Vulture in Douglas Co., Oregon. Oologist 21:55.

Pennycuick, C. J. 1968. Power requirements for horizontal flight in the Pigeon Columbia livia. Journal of Experimental Biology 49:527–55.

Pennycuick, C. J. 1969. The mechanics of bird migration. Ibis 111:525–56.

Pennycuick, C. J. 1972. Soaring behavior and performance of some East African birds, observed from a motor-glider. Ibis 114:178–218.

Pennycuick, C. J., and K. D. Scholey. 1984. Flight behavior of Andean Condors Vultur gryphys and Turkey Vultures Cathartes aura around the Paracas Peninsula, Peru. Ibis 126:253–56.

Peterson, L. 1990. The Story of the Sechelt Nation. Harbour Publishing: Madeira Park, British Columbia.

Pettitt, G. A. 1950. The Quileute of La Push 1775–1945. Anthropological Records 14:1–128.

Phillips, W. E. 1976. The Conservation of the California Tule Elk: A Socioeconomic Study of a Survival Problem. University of Alberta Press: Edmonton.

Pitelka, F. A. 1981. The Condor case: An uphill struggle in a downhill crush. Auk 98:634–35.

Poesch, J. 1961. Titian Ramsay Peale, 1799–1885; and his Journals of the Wilkes Expedition. American Philosophical Society: Philadelphia.

Pomeroy, E. S. 1965. The Pacific Slope: A History of California, Oregon, Washington, Idaho, Utah, and Nevada. University of Nevada Press: Reno.

Porter, R. P., H. Gannett, and W. P. Jones. 1882. The West: From the Census of 1880, a History of the Industrial, Commercial, Social, and Political Development of the States and Territories of the West from 1800 to 1880. Rand, McNally & Company: Chicago.

Prakash, V., D. J. Pain, A. A. Cunningham, P. F. Donald, N. Prakash, A. Verma, R. Gargi, S. Sivakumar, and A. R. Rahmani. 2003. Catastrophic collapse of Indian White-backed *Gyps bengalensis* and Long-billed *Gyps indicus* Vulture populations. Biological Conservation 109:381–90.

Prather, I. D., R. N. Conner, and C. S. Adkisson. 1976. Unusually large vulture roost in Virginia. Wilson Bulletin 88:667–68.

Prior, K. A. 1990. Turkey Vulture food habits in southern Ontario. Wilson Bulletin 102:706–10.

Puget Sound Argus (Port Townsend, Washington). 1882. Whale story of the Makahs. 27 October.

Putnam, R. (with transcript and notes by S. Hargreaves). 1928. The letters of Roselle Putnam. Oregon Historical Quarterly 29:242–64.

Ragir, S. 1972. The early horizon in central California prehistory. Contributions of the University of California Archaeological Research Facility 15:1–158.

Rajala, R. A. 1998. Clearcutting the Pacific Rainforest: Production, Science, and Regulation. University of British Columbia Press: Vancouver.

Reagan, A. B. 1909. Some notes on the Olympic Peninsula, Washington. Transactions of the Kansas Academy of Science 22:131–238.

Reagan, A. B. 1917. Archaeological notes on western Washington and adjacent British Columbia. Proceedings of the California Academy of Sciences 7:1–32.

Reed, G. W., and R. Gaines (editors). 1949. The Journals, Drawings and Other Papers of J. Goldsborough Bruff, April 2, 1849–July 20, 1851. Columbia University Press: New York.

Reid, M. (editor). 1869. The Vultures of America: A Monographic Sketch of these Foul-beaked Birds. Mayne Reid's Magazine, Onward, April 1869:281–87; May 1869:371–87.

Revere, J. W. 1849. A Tour of Duty in California, Including a Description of the Gold Region: and an Account of the Voyage Around Cape Horn; With Notices of Lower California, the Gulf and Pacific Coasts, and the Principal Events Attending the Conquest of the Californians. C. S. Francis & Co.: New York.

Rhoads, S. 1893. The birds observed in British Columbia and Washington during spring and summer, 1892–1893. Proceedings of the Academy of Natural Sciences of Philadelphia 45:21–65.

Rich, E. E. (editor). 1943. The letters of John McLoughlin from Fort Vancouver to the Governor and Committee, Second Series, 1839–44. Champlain Society: Toronto, Ontario.

Ricklefs, R. E. (editor). 1978. Report of the Advisory Panel on the California Condor. Audubon Conservation Report 6. National Audubon Society: New York.

Rideout, B. A., I. Stalis, R. Papendick, A. Pessier, B. Puschner, M. E. Finkelstein, D. R. Smith, M. Johnson, M. Mace, R. Stroud, J. Brandt, J. Burnett, C. Parish, J. Petterson, C. Witte, C. Stringfield, K. Orr, J. Zuba, M. Wallace, and J. Grantham. 2012. Patterns of mortality in free-ranging California Condors (*Gymnogyps californianus*). Journal of Wildlife Diseases 48:95–112.

Rising, H. G. 1899. Capture of a California Condor. Bulletin of the Cooper Ornithological Club 1:25–26.

Rodríguez, J. P. 2002. Range contraction in declining North American bird populations. Ecological Applications 12:238–48.

Ross, A. 1855. The Fur Hunters of the Far West; A Narrative of Adventures in the Oregon and Rocky Mountains. Smith, Elder and Co.: London.

Rowe, S. P., and T. Gallion. 1996. Fall migration of Turkey Vultures and raptors through the southern Sierra Nevada, California. Western Birds 27:48–53.

Ruxton, G. D., and D. C. Houston. 2004. Obligate vertebrate scavengers must be large soaring fliers. Journal of Theoretical Biology 228:431–36.

Sage, R. B. 1846. Scenes in the Rocky Mountains, and in Oregon, California, New Mexico, Texas, and the Great Prairies; or Notes by the Way, During an Excursion of Three Years, with a Description of the Countries Passed Through, Including Their Geography, Geology, Resources, Present Condition, and the Different Nations Inhabiting Them. By a New Englander. Cary & Hart: Philadelphia.

Sapir, E. 2001. The Collected Works of Edward Sapir. P. Sapir (editor). Mouton de Gruyter: Berlin.

Sapir, E., and L. Spier. 1943. Notes on the culture of the Yana. Anthropological Records 3:239–98.

Sapir, E., and M. Swadesh. 1978. Nootka Texts: Tales and Ethnological Narratives, with Grammatical Notes and Lexical Materials. AMS Press: New York.

Sawyer, J. O. 2006. Northwest California: A Natural History. University of California Press: Berkeley and Los Angeles.

Sawyer, R. W. 1932. Abbot railroad surveys, 1855. Oregon Historical Quarterly 33:1–24, 115–35.

Scammon, C. M. 1874. The Marine Mammals of the Northwestern Coast of North America. John H. Carmany and Company: San Francisco.

Schaeffer, C. E. 1951. Was the California Condor known to the Blackfoot Indians? Journal of the Washington Academy of Sciences 41:181–91.

Schaeffle, E. 1915. Fish and game: One of California's great resources. California's Magazine 1:159–68.

Schafer, J. 1909. A History of the Pacific Northwest. MacMillan Company: New York.

Scheffer, T. H. 1928. Status of the seal and sea-lion on our Northwest coast. Journal of Mammalogy 9:10–16.

Schenck, W. E., and E. J. Dawson. 1929. Archaeology of the northern San Joaquin Valley. University of California Publications in American Archaeology and Ethnology 25:289–413.

Schlick, M. D. 1994. Columbia River Basketry: Gift of the Ancestors, Gift of the Earth. University of Washington Press: Seattle.

Schoning, R. W., T. R. Merrell Jr., D. R. Johnson. 1951. The Indian Dip Net Fishery at Celilo Falls on the Columbia River. Oregon Fish Commission Contribution 17: Portland, Oregon.

Schrepfer, S. R. 1983. The Fight to Save the Redwoods: A History of Environmental Reform, 1917–1978. University of Wisconsin Press: Madison.

Schwartz, E. A. 1997. The Rogue River Indian War and Its Aftermath, 1850–1980. University of Oklahoma Press: Norman.

Sclater, P. L. 1866. Living California Vulture received in London. Proceedings of the Zoological Society of London 13:366.

Scott, C. D. 1936a. Are Condors extinct in Lower California? Condor 38:41–42.

Scott, C. D. 1936b. Who killed the Condors? Nature Magazine 28:368–70.

Scouler, J. 1905. Dr. John Scouler's journal of a voyage to N. W. America [1824–'25–'26.] III. Departing visit to the Columbia on the return from the voyage to the north and homeward bound. Quarterly of the Oregon Historical Society 6:276–85.

Seabough, S. 1880. Gold Lake. Search for a mythical bonanza in the Sierras. *Chronicle* (San Francisco newspaper) 21 November.

Seddon, P. J., D. P. Armstrong, and R. F. Maloney. 2007. Developing the science of reintroduction biology. Conservation Biology 21:303–12.

Seibold, I., and A. Helbig. 1995. Evolutionary history of New and Old World Vultures inferred from nucleotide sequences of the mitochondrial cytochrome *b* gene. Philosophical Transactions of the Royal Society B: Biological Sciences 350:163–78.

Sharp, B. E. 2012. The California Condor in northwestern North America. Western Birds 43:54–89.

Sheldon, H. H. 1939. What price condor? Field and Stream 44:22–23, 61–62.

Shepherdson, D., A. Warner, J. Hartline, S. St. Michael, D. Moen, and R. Walker. 2007. Oregon condor reintroduction: A feasibility proposal. Oregon Zoo unpublished report. 6 pp.

Shufeldt, R. W. 1900. Chapters on the Natural History of the United States. Studer Brothers Publishers: New York.

Shultz, S., H. S. Baral, S. Charman, A. A. Cunningham, D. Das, G. R. Ghalsasi, M. S. Goudar, R. E. Green, A. Jones, P. Nighot, D. J. Pain, and V. Prakash. 2004. Diclofenac poisoning is widespread in declining vulture populations across the Indian subcontinent. Proceedings of the Royal Society of London: Biological Sciences 271:S458–460.

Simons, D. D. 1983. Interactions between California Condors and humans in prehistoric far western North America. Pages 470–94 *in* S. R. Wilbur and

J. A. Jackson (editors), Vulture Biology and Management. University of California Press: Berkeley.

Sinclair, W. J. 1904. The exploration of Potter Creek Cave. University of California Publications in American Archaeology and Ethnology 2:1–27.

Smith, C. R. 2006. Bigger is better: The role of whales as detritus in marine ecosystems. Pages 286–300 (chap. 22) *in* J. A. Estes, D. P. Demaster, D. F. Doak, T. M. Williams, and R. L. Brownell Jr. (editors), Whales, Whaling, and Ocean Ecosystems. University of California Press: Berkeley.

Smith, D. 1978. Condor Journal: The History, Mythology and Reality of the California Condor. Capra Press: Santa Barbara, California.

Smith, D., and R. Easton. 1964. California Condor, Vanishing American: A Study of an Ancient Symbolic Giant of the Sky. McNally and Loftin: Charlotte, North Carolina/Santa Barbara, California.

Smith, F. J. 1916. Occurrence of the condor in Humboldt County. Condor 18:205.

Snyder, N. F. R., S. R. Derrickson, S. R. Beissinger, J. W. Wiley, T. B. Smith, and W. D. Toone. 1996. Limitations of captive breeding in endangered species recovery. Conservation Biology 10:338–48.

Snyder, N. F. R., and J. A. Hamber. 1985. Replacement-clutching and annual nesting of California Condors. Condor 87:374–78.

Snyder, N. F. R., and E. V. Johnson. 1985. Photographic censusing of the 1982–1983 California Condor population. Condor 87:1–13.

Snyder, N. F. R., and V. J. Meretsky. 2003. California Condors and DDE: A re-evaluation. Ibis 145:136–51.

Snyder, N. F. R., R. R. Ramey, and F. C. Sibley. 1986. Nest-site biology of the California Condor. Condor 88:228–41.

Snyder, N. F. R., and N. J. Schmitt. 2002. California Condor (*Gymnogyps californianus*). *In* A. Poole (editor), The Birds of North America Online. Cornell Lab of Ornithology: Ithaca, New York, http://bna.birds.cornell.edu/bna/species/610doi:10.2173/bna.610.

Snyder, N. F. R., and H. A. Snyder. 1989. Biology and conservation of the California Condor. Current Ornithology, 6:175–267.

Snyder, N. F. R., and H. A. Snyder. 2000. The California Condor: A Saga of Natural History and Conservation. Academic Press: San Diego, California.

Snyder, N. F. R., and H. A. Snyder. 2005. Introduction to the California Condor. University of California Press: Berkeley.

Soulé, M. E. 1983. What do we really know about extinction? Pages 111–24 *in* C. M. Schonewald-Cox, S. M. Chambers, B. MacBryde, and W. L. Thomas (editors), Genetics and conservation: A reference for managing wild animal and plant populations. Benjamin/Cummings Publishing Company: London, United Kingdom.

Stager, K. E. 1967. Avian olfaction. American Zoologist 7:415–20.

Stauber, E., N. Finch, P. A. Talcott, and J. M. Gay. 2010. Lead poisoning of Bald (*Haliaeetus leucocephalus*) and Golden (*Aquila chrysaetos*) Eagles in the US

Inland Pacific Northwest Region—an 18-year retrospective study: 1991–2008. Journal of Avian Medicine and Surgery 24:279–87.

Steadman, D. W., J. Arroyo-Cabrales, E. Johnson, and A. F. Guzman. 1994. New information on the late Pleistocene birds from San Josecito Cave, Nuevo León, Mexico. Condor 96:577–89.

Steadman, D. W., and N. G. Miller. 1987. California Condor associated with spruce-jack pine woodland in the late Pleistocene of New York. Quaternary Research 28:415–426.

Stendell, R. C. 1980. Dietary exposure of kestrels to lead. Journal of Wildlife Management 44:527–30.

Stephens, F. 1899. Lassoing a California Vulture. Bulletin of the Cooper Ornithological Club 1:88.

Stewart, P. A. 1977. Migratory movements and mortality rate of Turkey Vultures. Bird-Banding 48:122–24.

Stillman, J. D. B. 1967. The Gold Rush Letters of J. D. B. Stillman, with an introduction by Kenneth Johnson. Lewis Osborne: Palo Alto, California.

Stoms, D. M., F. W. Davis, C. B. Cogan, M. O. Painho, B. W. Duncan, J. Scepan, and J. M. Scott. 1993. Geographic analysis of California Condor sighting data. Conservation Biology 7:148–59.

Stone, W. 1916. Philadelphia to the coast in early days, and the development of Western ornithology prior to 1850. Condor 18:3–14.

Streator, C. P. 1888. Notes on the California Condor. Oologist and Ornithologist 13:30.

Sturtevant, W. C. 1967. Early Indian Tribes, Culture Areas, and Linguistic Stocks [map]. Smithsonian Institution, National Atlas, 1:7,500,000.

Suárez, W. 2000. Contribución al conocimiento del estatus generic del condor extinto (Ciconiiformes: Vulturidae) del Cuaternario Cubano. Ornithologica Neotropical 11:109–22.

Sullivan, M. S. 1934. The Travels of Jedediah Smith: A Documentary Outline, Including the Journal of the Great American Pathfinder. University of Nebraska Press: Lincoln.

Sumich, J. L. 1984. Gray whales along the Oregon coast in summer, 1977–1980. The Murrelet 65:33–40.

Suttles, W. (editor). 1990. Handbook of North American Indians, Northwest Coast, vol. 7. Smithsonian Institution: Washington, DC.

Swan, J. G. 1870. The Indians of Cape Flattery, at the Entrance to the Strait of Fuca, Washington Territory. Smithsonian Contributions to Knowledge 220. Smithsonian Institution: Washington, DC.

Swan, J. G. 1887. The fur seal industry of Cape Flattery, Washington Territory. Pages 393–400 in G. B. Goode (editor), Fisheries and Fishery Industries of the United States, vols. 7 and 8. Government Printing Office: Washington, DC.

Swanton, J. R. 1909. Tlingit Myths and Texts. Smithsonian Institution Bureau of American Ethnology, Bulletin 39. Government Printing Office: Washington, DC.

Syverson, V. J., and D. R. Prothero. 2010. Evolutionary patterns in late Quaternary California Condors. PalArch's Journal of Vertebrate Palaeontology 7:1–18.

Szulkin, M., and B. C. Sheldon. 2008. Dispersal as a means of inbreeding avoidance in a wild bird population. Proceedings of the Royal Society B 275:703–11.

Taylor, A. S. 1859. The great condor of California. Hutchings' California Magazine 37:17–22.

Taylor, R. E. 1975. UCR Radiocarbon Dates II. Radiocarbon 17:396–406.

Taylor, W. P. 1916. The conservation of the native fauna. Scientific Monthly 3:399–409.

Tear, T. H., P. Kareiva, P. L. Angermeier, P. Comer, B. Czech, R. Kautz, L. Landon, D. Mehlman, K. Murphy, M. Ruckelshaus, J. M. Scott, and G. Wilhere. 2005. How much is enough? The recurrent problem of setting measurable objectives in conservation. BioScience 55:835–49.

The Nature Conservancy, in litt. 2007. A proposal to explore the establishment of a California Condor (re)introduction program in the Lassen Foothills Project Area, California. Letter from Simon Avery (Dye Creek Preserve Manager) to Mark Weitzel (Hopper Mountain National Wildlife Refuge Project Leader). March 1, 2007.

Thiollay, J.-M. 2006. The decline of raptors in West Africa: Long-term assessment and the role of protected areas. Ibis 148:240–54.

Tolmie, W. F. 1963. The Journals of William Fraser Tolmie, Physician and Fur Trader. Mitchell Press Limited: Vancouver, British Columbia.

Tønnessen, J. N., and A. O. Johnsen. 1982. The History of Modern Whaling (translated from Norwegian by R. I. Christophersen). University of California Press: Berkeley and Los Angeles.

Tower, W. S. 1907. A History of the American Whale Fishery. Publications of the University of Pennsylvania, Series in Political Economy and Public Law 20. John C. Winston Co.: Philadelphia.

Townsend, C. H. 1887. Field-notes on the mammals, birds and reptiles of northern California. Proceedings of the United States National Museum 10:159–241.

Townsend, J. K. 1839. Narrative of a Journey Across the Rocky Mountains, to the Columbia River, and a Visit to the Sandwich Islands, Chili, &c, with a Scientific Appendix. Henry Perkins: Philadelphia.

Townsend, J. K. 1848. Popular monograph of the Accipitrine birds of N.A.—No. II. The Literary Record and Journal of the Linnaean Association of Pennsylvania College 4:249–55, 265–72.

Tribune (Chicago newspaper). 1872. A Vulture Eagle, measuring upward of nine feet from tip to tip, has been killed in Mendocino County, Cal. February 24.

Tyler, J. G. 1918. Underappreciated friends. California Fish and Game 4:26–29.

Uhle, M. 1907. The Emeryville Shellmound. University of California Publications in American Archaeology and Ethnology 7:1–84 + plates.

Urios, V., P. López-López, R. Limiñana, and A. Godino. 2010. Ranging behaviour of a juvenile Bearded Vulture (*Gypaetus barbatus meridionalis*) in South Africa revealed by GPS satellite telemetry. Ornis Fennica 87:114–18.

US Department of the Interior. 1853. The Seventh Census of the United States: 1850. Robert Armstrong, Public Printer: Washington, DC.

US Department of the Interior. 1864. Agriculture of the United States in 1860; Compiled from the Original Returns of the Eighth Census. Government Printing Office: Washington, DC.

US Department of the Interior. 1895. Report on the Statistics of Agriculture in the United States at the Eleventh Census: 1890. Government Printing Office: Washington, DC.

US Department of the Interior. 1902. Census Reports, vol. 5: Twelfth Census of the United States, Taken in the Year 1900; Agriculture, part 1: Farms, Live Stock, and Animal Products. United States Census Office: Washington, DC.

USFWS (US Fish and Wildlife Service). 1975. California Condor Recovery Plan. US Fish and Wildlife Service: Portland, Oregon.

USFWS (US Fish and Wildlife Service). 1979. Recommendations for Implementing the California Condor Contingency Plan. February 23, 1979, unpublished report. Washington, DC.

USFWS (US Fish and Wildlife Service). 1980. California Condor Recovery Plan (first revision). US Fish and Wildlife Service: Portland, Oregon.

USFWS (US Fish and Wildlife Service). 1984. California Condor Recovery Plan (second revision). US Fish and Wildlife Service: Portland, Oregon.

USFWS (US Fish and Wildlife Service). 1996a. Endangered and threatened wildlife and plants: Establishment of a nonessential experimental population of California Condors in northern Arizona; Final Rule. Federal Register 61:54043–60.

USFWS (US Fish and Wildlife Service). 1996b. California Condor Recovery Plan (third revision). US Fish and Wildlife Service: Portland, Oregon.

Verner, J. 1978. California Condors: Status of the Recovery Effort. Pacific Southwest Forest and Range Experiment Station General Technical Report PSW-28. Forest Service, US Department of Agriculture: Berkeley, California.

Victor, F. F. 1872. All Over Oregon and Washington: Observations on the Country, Its Scenery, Soil, Climate, Resources, and Improvements, With an Outline of its Early History, and Remarks on its Geology, Botany, Mineralogy, Etc. Also Hints to Immigrants and Travelers Concerning Routes, the Cost of Travel, The Price of Land, Etc. John H. Carmany & Co.: San Francisco.

Virani, M. Z., C. Kendall, P. Njoroge, and S. Thomsett. 2011. Major declines in the abundance of vultures and other scavenging raptors in and around the Masai Mara ecosystem, Kenya. Biological Conservation 144:746–52.

Walker, D. E., Jr. 1997. The Yakama system of trade and exchange. Northwest Anthropological Research Notes 31:71–95.

Wallace, M. 1989. Andean Condor experimental releases to enhance California Condor recovery. Endangered Species Update 6 (March/April):1–4.

Wallace, W. J., and D. J. Lathrap. 1959. Ceremonial bird burials in San Francisco Bay shellmounds. American Antiquity 25:262–64.

Wallace, W. J., and D. J. Lathrap. 1975. West Berkeley (CA-Ala-307): A culturally stratified shellmound on the east shore of San Francisco Bay. Contributions of the University of California Archaeological Research Facility 29:1–64 + plates.

Walter, J. 2006. The Guns That Won the West: Firearms on the American Frontier, 1848–1898. Greenhill Books/Lionel Leventhal, Ltd. and MBI Publishing: London and St. Paul, Minnesota.

Walters, J. R, S. R. Derrickson, D. M. Fry, S. M. Haig, J. M. Marzluff, and J. M. Wunderle Jr. 2010. Status of the California Condor (Gymnogyps californianus) and efforts to achieve its recovery. Auk 127:969–1001.

Waterman, T. T. 1920. The whaling equipment of the Makah Indians. University of Washington Publications in Anthropology 1:1–67.

Webb, R. L. 1998. On the Northwest: Commercial Whaling in the Pacific Northwest 1790–1967. University of British Columbia Press: Vancouver.

Wetmore, A. 1931a. The California Condor in New Mexico. Condor 33:76–77.

Wetmore, A. 1931b. The Avifauna of the Pleistocene in Florida. Smithsonian Miscellaneous Collections 85(2). Smithsonian Institution: Washington, DC.

Whelan, T. 1918. The American Rifle: A Treatise, a Text Book, and a Book of Practical Instruction in the Use of the Rifle. Century Co.: New York.

White, T. H., Jr., N. J. Collar, R. J. Moorhouse, V. Sanz, E. D. Stolen, and D. J. Brightsmith. 2012. Psittacine reintroductions: Common denominators of success. Biological Conservation 148:106–15.

Wickersham, J. 1896. Some North-west burial customs. American Antiquarian 18:204–6.

Wiemeyer, S. N., J. M. Scott, M. P. Anderson, P. H. Bloom, and C. J. Stafford. 1988. Environmental contaminants in California Condors. Journal of Wildlife Management 52:238–47.

Wilbur, M. E. (editor). 1941. A Pioneer at Sutter's Fort, 1846–1850; the Adventures of Heinrich Lienhard. The Calafia Society: Los Angeles.

Wilbur, S. R. 1972. Food Resources of the California Condor. US Fish and Wildlife Service Administrative Report. Ojai, California. September 1972.

Wilbur, S. R. 1973. The California Condor in the Pacific Northwest. Auk 90:196–98.

Wilbur, S. R. 1978. The California Condor, 1966–76: A Look at its Past and Future. North American Fauna 72:1–136.

Wilbur, S. R. 1980. Estimating the size and trend of the California Condor population, 1965–1978. California Fish and Game 66:40–48.

Wilbur, S. R. 2004. Condor Tales: What I Learned in Twelve Years with the Big Birds. Symbios: Gresham, Oregon.

Wilcox, W. A. 1895. Fisheries of the Pacific Coast. Pages 139–304 *in* US Commission of Fish and Fisheries, part 19. Report of the Commissioner for the Year Ending June 30, 1893. Government Printing Office: Washington DC.

Wild, E. W. 1898. Shooting sea lions. Recreation 8:9–10.

Wilkes, C. 1849. Western America, Including California and Oregon, with Maps of Those Regions, and of "The Sacramento Valley." From Actual Surveys. Lea and Blanchard: Philadelphia.

Willamette Farmer (Salem, Oregon). 1872. Highland Farmers' Club. 6 July.

Willamette Farmer (Salem, Oregon). 1887. State and territorial news. 25 March.

Wink, M. 1995. Phylogeny of Old and New World vultures (Aves: Accipitridae and Cathartidae) inferred from nucleotide sequences of the mitochondrial cytochrome b gene. Zeitschrift fur Naturforschung C 50:868–82.

Winther, O. O. 1950 (reprinted 1969). The Old Oregon Country: A History of Frontier Trade, Transportation, and Travel. University of Nebraska Press: Lincoln.

Wobeser, G., T. Bollinger, F. A. Leighton, B. Blakley, and P. Mineau. 2004. Secondary poisoning of eagles following intentional poisoning of coyotes with anticholinesterase pesticides in western Canada. Journal of Wildlife Diseases 40:163–72.

Wood, R. L. 1995. The Land that Slept Late: The Olympic Mountains in Legend and History. The Mountaineers: Seattle.

Woods, C. P., W. R. Heinrich, S. C. Farry, C. N. Parish, S. A. H. Osborn, and T. J. Cade. 2007. Survival and reproduction of California Condors released in Arizona. Pages 57–78 *in* A. Mee and L. S. Hall (editors), California Condors in the 21st Century. Nuttall Ornithological Club: Cambridge, Massachusetts; American Ornithologists' Union: Washington, DC.

Woodwell, G. M., P. P. Craig, and H. A. Johnson. 1971. DDT in the biosphere: Where does it go? Science 174:1101–7.

Wray, J. (editor). 2002. Native Peoples of the Olympic Peninsula: Who We Are. University of Oklahoma Press: Norman.

Xirouchakis, S. M., and N. Poulakakis. 2008. Biometrics, sexual dimorphism and gender determination of Griffon Vultures *Gyps fulvus* from Crete. Ardea 96:91–98.

Yoshiyama, R. M., F. W. Fisher, and P. B. Moyle. 1998. Historical abundance and decline of chinook salmon in the Central Valley region of California. North American Journal of Fisheries Management 18:487–521.

Young, F. G. 1904. Literary remains of David Douglas, botanist of the Oregon Country, editorial prefatory notes. Quarterly of the Oregon Historical Society 5:215–22.

Yurok Tribe. 2007. Yurok Tribe condor release initiative: Linking traditional knowledge and culture with modern habitat restoration and species conservation. Proposal submitted October 1, 2007, to the US Fish and Wildlife Service, Sacramento, California, for a 2008 Tribal Wildlife Grant.

Appendix

Annotated historical records of California Condor observations in the Pacific Northwest. Numbers correspond to the points on figure 15.

Record Number: 1

Location: Near the confluence of the Wind River and Columbia River, Skamania County, WA

Observer(s): Lewis and Clark party [October 1805 weather reports in Voorhis 4 and Codex I, Clark's journal (Oct. 30)]

Date: 28–30 October 1805

Observation Type: Observation

Reference(s): Lewis et al. 2002 (30 October 1805)[1]

Quotation(s): *28 October (weather report in Voorhis No. 4)*: "first Vulture of the Columbia Seen to day"

29 October (weather report in Codex I): "rained moderately all day. Saw the first large Buzzard or Voultur of the Columbia."[2]

30 October (Clark): "Scattered about in the river, this day we Saw Some fiew of the large Buzzard[.] Capt. Lewis Shot at one, those Buzzards are much larger than any other of ther Spece or the largest Eagle white under part of their wings &c."[3]

---

1 In addition to Lewis and Clark's journals, several members of the expedition kept journals, some entries of which are similar or the same for a given day. Apparently, copying journal entries from each other was not uncommon. Our records include all instances where the expedition mentioned California Condors, although journal entries that provided no additional details to those of Lewis or Clark were left out. Readers wishing to see all of the journal entries are directed to Lewis et al. (2002).

2 In Voorhis No. 4 for 29 October Clark writes, "I shot at a vulture" (Lewis et al. 2002).

3 Lewis makes no mention of this observation in his journal for that day.

Record Number: 2

Location: Cape Disappointment, WA

Observer(s): Lewis and Clark party [Clark's journal]

Date: 18 November 1805

Observation Type: Killed; specimen preserved but lost

Reference(s): Lewis et al. 2002 (18 November 1805)

Quotation(s):

Clark's journal, 18 November: "Rubin Fields Killed a Buzzard of the large Kind near the meat of the whale we Saw: W. 25 lb. measured from the tips of the wings across 9½ feet, from the point of the Bill to the end of the tail 3 feet 10¼ inches, middle toe 5½ inches, toe nale 1 inch & 3½ lines, wing feather 2½ feet long & 1 inch 5 lines diamiter tale feathers 14½ inches, and the *head* is 6½ inches including the beak."[4]

Gass's journal, 20 November: "Wednesday [November] 20th.—We had a fine clear morning . . . . At four o'clock in the afternoon, Capt. Clarke and his party returned to camp, and had killed a deer and some brants. They had been about 10 miles north of the cape [Cape Disappointment], and found the country along the seashore level with spruce-pine timber, and some prairies and ponds of water. They killed a remarkably large buzzard, of a species different from any I had seen. It was 9 feet across the wings, and 3 feet 10 inches from the bill to the tail."

Record Number: 3

Location: Near Astoria, OR

Observer(s): Lewis and Clark party [Clark's journal]

Date: 29–30 November 1805

Observation Type: Observation

Reference(s): Lewis et al. 2002 (29–30 November 1805)

---

4 Nicholas Biddle (first editor of the Lewis and Clark journals) noted that the head of this specimen was in Peale's Museum (Lewis et al. 2002). Charles Willson Peale, a noted painter, was the proprietor of Peale's Museum in Philadelphia, the only natural history museum in the country at that time. The museum was originally housed in present-day Independence Hall but was later moved to Baltimore, Maryland. The current location of the specimen is unknown.

Quotation(s): *29 November*: "The Shore below the point at our Camp is formed of butifull pebble of various colours. I observe but fiew birds of the Small kind, great numbers of wild fowls of Various kinds, the large Buzzard with white wings, grey and bald eagle's."

30 November: "Some rain and hail with intervales of fair weather for 1 and 2 hours dureing the night and untill 9 oClock this morning at which time it Cleared up fair and the Sun Shown, I Send 5 men in a Canoe in the Deep bend above the Peninsulear to hunt fowles, & 2 men in the thick woods to hunt Elk[.] had all our wet articles dried & the men all employed dressing their Skins, I observe but few birds in this Countrey of the Small kinds—great numbers of wild fowl, The large Buzzard with white under their wings Grey & Bald eagle large red tailed hawk, ravins, Crows, & a small brown bird which is found about logs &c. but fiew small hawks or other smaller birds to be seen at this time."[5]

Record Number: 4

Location: Near Fort Clatsop, OR

Observer(s): Lewis and Clark party [Lewis's journal]

Date: 3 January 1806

Observation Type: Observation

Reference(s): Lewis et al. 2002 (3 January 1806)

Quotation(s): "the bald Eagle and the beatifull Buzzard of the columbia still continue with us."

Record Number: 5

Location: Youngs River, Clatsop County, OR

Observer(s): Lewis and Clark party [Clark's journal][6]

Date: 16 February 1806

Observation Type: Live-captured, then killed

Reference(s): Lewis et al. 2002 (16–17 February 1806)

---

5 Lewis's journal for this day indicates he saw "a great abundance of fowls" but does not specifically list the California Condor.

6 Lewis's journal includes a similar entry on 17 February 1806, along with a drawing of the head of the California Condor (figure 16).

Quotation(s): "Shannon an[d] Labiesh brought in to us to day a Buzzard or *Vulture* of the Columbia which they had wounded and taken alive. I believe this to be the largest Bird of North America . . . we have Seen it feeding on the remains of the whale and other fish which have been thrown up by the waves on the Sea Coast. these I believe constitute their principal food, but I have no doubt but that they also feed on flesh. we did not meet with this bird un[t]ille we had decended the Columbia below the great falls [Celilo Falls]; and have found them more abundant below tide water than above. this is the Same Species of Bird which R. Field killed on the 18th of Novr. last and which is noticed on that day tho' not fully discribed then I thought this of the Buzzard Specis. I now believe that this bird is reather of the Vulture genus than any other, tho' it wants Some of their characteristics particularly the hair on the neck, and the feathers on the legs. This is a handsom bird at a little distance. it's neck is proportionably longer than those of the Hawks or Eagle. . . . Shannon and Labiesh informed us that when he approached this Vulture after wounding it, that it made a loud noise very much like the barking of a Dog."

Record Number: 6

Location: Near Fort Clatsop, OR

Observer(s): Lewis and Clark party [Gass's journal]

Date: 15 March 1806

Observation Type: Killed (x2)

Reference(s): Lewis et al. 2002 (16 March 1806)

Quotation(s): "Yesterday [15 March 1806] while I was absent getting our meat home, one of the hunters killed two vultures, the largest fowls I had ever seen. I never saw any such as these except on the Columbia river and the seacoast."[7]

Record Number: 7

Location: North end of Deer Island, Columbia County, OR

Observer(s): Lewis and Clark party [Clark's journal, Lewis's journal]

Date: 28 March 1806

---

7 There is no mention of this incident in either Lewis's or Clark's journal; perhaps this is not surprising given that they had already described the species in detail and they had long entries for other species around this time.

Observation Type: Observation

Reference(s): Lewis et al. 2002 (28 March 1806)

Quotation(s): *Clark's journal*: "The men who had been Sent after the deer returned with four only, the other 4 haveing been eaten entirely by the Voulturs[8] except the Skin."

Lewis's journal: "The men who had been sent after the deer returned and brought in the remnent which the Vultures and Eagles had left us; these birds had devoured 4 deer in the course of a few hours. the party killed and brought in three other deer a goose some ducks and an Eagle. Drewyer also killed a tiger cat. Joseph Fields informed me that the Vultures had draged a large buck which he had killed about 30 yards, had skined it and broken the back bone."

Record Number: 8

Location: Near Rooster Rock State Park, OR

Observer(s): Lewis and Clark party [Lewis's journal, Clark's journal]

Date: 5–6 April 1806

Observation Type: Killed

Reference(s): Lewis et al. 2002 (5–6 April 1806)

Quotation(s): Lewis's journal, 5 April 1806: "we saw the martin, small gees, the small speckled woodpecker with a white back, the Blue crested Corvus, ravens, crows, eagles Vultures and hawks."

Clark's journal, 6 April 1806: "Jos: Field killed a vulture of that Speces already described."[9]

Record Number: 9

Location: Lower end of Hamilton Island, WA

Observer(s): Alexander Henry and David Thompson

Date: 19 January 1814

Observation Type: Observation

---

8 Lewis and Clark's party did not see Turkey Vultures west of the Rocky Mountains until 9 April 1806.

9 The only vulture species that the party had described in their journals to this point was the California Condor. Lewis and Clark's party did not see Turkey Vultures west of the Rocky Mountains until 9 April 1806.

Reference(s): Coues 1897:808

Quotation(s): "Some extraordinarily large vultures [editorial note by Coues: *Pseudogryphus californianus*] were hovering over camp."

Record Number: 10

Location: Above Willamette Falls, near Pudding River, OR

Observer(s): Alexander Henry and David Thompson

Date: January 1814

Observation Type: Observation

Reference(s): Coues 1897:817

Quotation(s): "I sent for the eight deer killed yesterday. The men brought in seven of them, one having been devoured by the vultures [editorial note by Coues: *Pseudogryphus californianus*]. These birds are uncommonly large and very troublesome to my hunters by destroying the meat, which, though well covered with pine branches, they contrive to uncover and devour."

Record Number: 11

Location: Idaho/Oregon border[10]

Observer(s): Donald McKenzie

Date: Fall 1818

Observation Type: Observation

Reference(s): Ross 1855:203

Quotation(s): "Eagles and vultures, of uncommon size, flew about the rivers."

Record Number: 12

Location: Near the confluence of the Cowlitz and Columbia Rivers, WA

Observer(s): John Scouler

Date: September 1825

---

10 Exact location unclear; somewhere in the mountainous terrain southeast of Walla Walla, Washington, probably near the Idaho border.

Observation Type: Killed; specimen 162189 in the National Museum of Natural History, Leiden, The Netherlands.

Reference(s): Scouler 1905:280

Quotation(s): *22 September 1825:* "This morning we breakfasted at the Kowlitch [Cowlitz] village & we were treated with much civility, although they were in a very unsettled state & were preparing for war in consequence of the circumstances formerly alluded to.

On arriving on board the ship much of my time was employed procuring & preserving birds. The incessant rains we experienced at the advanced period of the year rendered the accumulation of plants hopeless. The river at this season was beginning to abound in birds. I obtained specimens of *Pelecanus onocrotalus, Falco*—& a species of *Vultur*, which I think is nondescript. My birds were princip[a]lly obtained from the Indians, who would go through any fatigue for a bit of tobac[c]o."

Record Number: 13

Location: Near Fort Vancouver, WA

Observer(s): David Douglas

Date: Winter 1826

Observation Type: Killed[11]

Reference(s): Douglas 1914:154–55; Hall 1934:5

Quotation(s): *Douglas 1914, 154–55:* "On the Columbia there is a species of Buzzard, the largest of all birds here, the Swan excepted. I killed one of this very interesting bird, with buckshot, one of which passed through the head, which rendered it unfit for preserving; I regret it exceedingly, for I am confident it is not yet described. I have fired at them with every size of small shots at respectable distances without effect; seldom more than one or two are together. When they find a dead carcase or putrid animal matter, so gluttonous are they that they will eat until they can hardly walk and have been killed with a stick. They are of the same colour as the common small buzzard found in Canada, one of which was sent home last October. Beak and legs bright yellow. The feathers of the

---

11 This bird was preserved and eventually ended up in either the Institute Royal des Sciences Naturelles de Belgique (Brussels) or the Museum National d'Histoire Naturelle (Paris); the location of this specimen and the other specimen Douglas collected and preserved (record 17) cannot be assigned with certainty.

wing are highly prized by the Canadian voyageurs for making tobacco pipe-stems. I am shortly to try to take them in a baited steel-trap."

Hall 1934, 5: "When opportunity favoured I collected woods, and gathered Musci, &c., and from this time to March 20th I formed a tolerable collection of preserved animals and birds . . . [including] *Sacroramphos californica* [California Condor]."

Record Number: 14

Location: Between the Umpqua River and Willamette Valley, OR

Observer(s): David Douglas

Date: 10 or 11 October 1826

Observation Type: Observation

Reference(s): Douglas 1914:67

Quotation(s): "This morning [10 or 11 October 1826] we passed a hill of similar elevation and appearance to that we passed yesterday. Several species of *Clethra* were gathered—one in particular, *C. grandis* [*sic*], was very fine—and many birds of *Sacoramphos californica* [California Condor] and *Ortyx californica*, and two other species of great beauty were collected. This part of the time was rainy, ill-adapted for hunting. The last two days' march we descended the banks of Red Deer River, which empties itself into the River Arguilar or Umpqua, forty-three miles from the sea."

Record Number: 15

Location: Umpqua River and south, OR

Observer(s): David Douglas and Norman McLeod

Date: October 1826

Observation Type: Observation

Reference(s): Douglas 1914:216, 241

Quotation(s): *Douglas 1914, 216*: "The Large Buzzard, so common on the shores of the Columbia, is also plentiful here [Umpqua River region]; saw nine in one flock."

Douglas 1914, 241: "I think they [California Condors] migrate to the south, as great numbers were seen by myself on the Umpqua River, and south of it by Mr. McLeod, whom I accompanied."

Record Number: 16

Location: East of the Cascade Range, near the US-Canada border, WA

Observer(s): David Douglas

Date: 1826–1827

Observation Type: Observation

Reference(s): Douglas 1829:329

Quotation(s): "I have met with them [*Vultur californianus*] as far to the north as 49° N. Lat. [the Canadian border] in the summer and autumn months, but no where so abundantly as in the Columbian valley between the Grand Rapids [Celilo Falls] and the sea."[12]

Record Number: 17

Location: Near Fort Vancouver, WA

Observer(s): David Douglas

Date: 3 February 1827

Observation Type: Killed[13]

Reference(s): Barnston 1860:208; Fleming 1924:111–12; Douglas 1914:241

Quotation(s): *Barnston 1860, 208; Fleming 1924, 111–12:* "The Spring of 1827 was severe, and much snow had fallen.[14] The consequence was that many horses died at Fort Vancouver, and we were visited by the various species of beasts and birds of prey that abound in that country. Most conspicuous among these were the California vulture. This magnate of the air was ever hovering around, wheeling in successive circles for a time, then changing the wing as if wishing to describe the figure 8; the ends of the pinions, when near enough to be seen, hav-

---

12 Although Douglas reports here that condors occurred north to the Canadian border (during his travels to the east of the Cascade Range), he makes no specific mention of these sightings in his journals, which is odd given the details he provides for the condors he observed and collected west of the Cascades.

13 This bird was preserved and eventually ended up in either the Institute Royal des Sciences Naturelles de Belgique (Brussels) or the Museum National d'Histoire Naturelle (Paris); the location of this specimen and the other specimen Douglas collected and preserved (record 13) cannot be assigned with certainty.

14 Although Barnston writes "Spring," in the next paragraph he begins with "March of 1827 arrived," consistent with the February date given by Douglas.

ing a bend waving upwards, all his movements, whether of soaring or floating ascending or descending, were lines of beauty. In flight he is the most majestic bird I have seen. One morning a large specimen was brought into our square. . . . It has been frequently a matter of surprise how quickly these birds collect when a large animal dies. None may be seen in any direction, but in a few minutes after a horse or other large animal gives up the ghost they may be descried like specks in the æther, nearing by circles to their prey."

Douglas 1914, 241: "Killed a very large vulture, sex unknown . . . During the summer [they] are seen in great numbers on the woody part of the Columbia, from the ocean to the mountains of Lewis and Clarke's River [Columbia River], four hundred miles in the interior. In winter they are less abundant. . . . Feeds on all putrid animal matter and are so ravenous that they will eat until they are unable to fly. Are very shy: can rarely get near enough to kill them with buck-shot; readily taken with a steel trap."

Record Number: 18

Location: Klamath River area near present-day Humboldt-Del Norte county line, CA

Observer(s): Jedediah Smith

Date: 22 May 1828

Observation Type: Observation

Reference(s): Sullivan 1934:92

Quotation(s): "Among the animals I observed in the country [were] . . . Large & small Buzards, Crows, Ducks, Ravens, several kinds of hawks, Eagles and a few small birds."[15]

Record Number: 19

Location: Cowlitz River, near its confluence with the Columbia River, WA

Observer(s): William Fraser Tolmie

---

15 The fact that Jedediah Smith differentiates between large and small buzzards, hawks, and eagles, along with the location and time period, makes this a credible sighting. Many other observers subsequently found condors to be numerous in northern California in the mid-1800s.

Date: 21 May 1833

Observation Type: Observation

Reference(s): Tolmie 1963:185

Quotation(s): "At 11½ passed our yesterday's encampment & scared some large vultures & crows from their feast."[16]

Record Number: 20

Location: Cowlitz River, WA

Observer(s): William Fraser Tolmie

Date: 22 May 1833

Observation Type: Observation

Reference(s): Tolmie 1963:186

Quotation(s): "Arrived about 11 at a deserted Indian village & startled some large vultures, who hovering above at length perched on the neighbouring trees, awaiting our departure . . . fired twice at vultures."[17]

Record Number: 21

Location: Near Fort Vancouver, WA

Observer(s): John Kirk Townsend

Date: 1834–1836

Observation Type: Observation

Reference(s): Audubon 1840:13

Quotation(s): "I once saw two [California Condors] near Fort Vancouver feeding on the carcass of a pig that had died."

Record Number: 22

Location: Willamette Falls, OR

Observer(s): John Kirk Townsend

Date: April 1835

---

16 The fact that Tolmie describes "large" vultures, along with the known occurrence of condors along the lower Columbia River during this time period, makes this a credible sighting.

17 See note 16 above.

Observation Type: Killed; US National Museum specimen 78005 (juvenile)

Reference(s): J. Townsend 1848:265–67

Quotation(s): "As I gazed upon them [Turkey Vultures], interested in their graceful and easy motions, I heard a loud rustling sound over my head, which induced me to look upward; and there, to my inexpressible joy, soared the great Californian, seemingly intent upon watching the motions of his puny relatives below. Suddenly, while I watched, I saw him wheel, and down like an arrow he plunged, alighting upon an unfortunate Salmon which had just been cast, exhausted with his attempts to leap the falls, on the shore within a short distance. At that moment I fired, and the poor Vulture fell wounded, beside his still palpitating quarry."[18]

Record Number: 23

Location: Sacramento Valley or Sonoma County, CA

Observer(s): Ivan Gavrilovich Voznesenskii

Date: 1840–1841

Observation Type: Killed (x4); specimens preserved and shipped to the Russian Academy of Sciences, but two were subsequently lost. The other two are now at the Saint Petersburg Zoological Museum, Russia (specimens 1583 and 1584).[19]

Reference(s): Blomkvist 1972:100–170; Bates 1983:36–41; Alekseev 1987

Quotation(s): None

---

18 The story continues with Townsend swimming (naked) across the river and fighting with the condor, including throwing rocks and kicking sand in its eyes. He ultimately killed the bird when he "hit him fairly on the head with a large stone" and then pounced on him and dispatched him by severing the spine with a knife.

19 In addition to four condor specimens, Voznesenskii also obtained an entire condor skin used by the Central Miwok Indians as a dance costume, and a condor cape. The dance costume and condor cape are now housed in the Peter the Great Museum of Anthropology and Ethnography, Saint Petersburg, Russia. The precise location where Voznesenskii collected his specimens is unclear. However, we know that from 11 April 1841 to 5 September 1841 he was in the vicinity of Fort Ross, Sonoma County, California, where he and A. G. Rotchev collected many birds. In May and June 1841, he traveled the entire length of the Russian River, and on 16 June 1841 made the first ascent of Mount Saint Helena, Sonoma County, California.

Record Number: 24

Location: Plains of the Willamette Valley, OR

Observer(s): Titian Ramsay Peale

Date: September 1841

Observation Type: Observation

Reference(s): Peale 1848:58

Quotation(s): "This cannot be considered a common bird in Oregon; we first saw them on the plains of the Willamette River, but subsequently observed that they were much more numerous in California, from the fact that the carcasses of large animals are more abundant, which they certainly prefer to the dead fish on which they are obliged to feed in Oregon, and all the countries north of the Spanish settlements in California."

Record Number: 25

Location: Near Youngs River (creek) in the Umpqua Mountains, OR

Observer(s): Titian Ramsay Peale

Date: 24 September 1841

Observation Type: Observation

Reference(s): Poesch 1961:191

Quotation(s): "we saw today Goldenwing woodpeckers (red var.), Ravens, Crows, Stellers & Florida Jays, Californian Vultures, and a few Larks."

Record Number: 26

Location: North of Redding (along the Sacramento River), CA

Observer(s): Titian Ramsay Peale

Date: 5 October 1841

Observation Type: Observation

Reference(s): Poesch 1961:194

Quotation(s): "I saw two species of marmots, and several birds not seen before. Sev California Vultures, etc."

Record Number: 27

Location: Sacramento Valley (between Redding and Sacramento), CA

Observer(s): Titian Ramsay Peale

Date: 18 October 1841

Observation Type: Observation

Reference(s): Poesch 1961:195

Quotation(s): "Numbers of Californian Vultures, Turkey buzzards, and Ravens were assembled this morning to enjoy the feast we have prepared for them."

Record Number: 28

Location: Valley of Napa Creek, CA

Observer(s): James Clyman

Date: August 1845

Observation Type: Observation

Reference(s): Clyman 1926:137

Quotation(s): "Beside the raven and turk[e]y Buzzard of the states you see here the royal vulture [California Condor] in greate abundance frequently measureing Fourteen feet from the extremity of one wing to the ext[r]emity of the other."

Record Number: 29

Location: Napa County, CA

Observer(s): James Clyman

Date: 8 September 1845

Observation Type: Killed

Reference(s): Clyman 1926:138

Quotation(s): "Killed 5 Deer one large grissled Bear one wild cat and a Royal vulture [California Condor] this is the largest fowl I have yet seen measuring when full grown full 14 feet from the ext[r]emity of one wing to the ext[r]emity of the other. Like all the vulture tribe this fowl feeds on dead carcases but like the Bald Eagle prefers his meat fresh and unputrefied[.] they seem [to] hover over these mountains in greate numbers are never at the least fault for their prey but move directly and rapidly to the carcase cutting the wind with their wings and creating a Buzzing sound which may [be] heard at a miles distance and making one or two curves they immediately alight and commence glutting[.]"

Record Number: 30

Location: Between Sutter's Fort and Suisun Bay, CA

Observer(s): Heinrich Lienhard

Date: 1846

Observation Type: Observation

Reference(s): M. Wilbur 1941:42

Quotation(s): "I had seen large numbers of vultures, turkey buzzards, ravens, crows, and magpies perched on a sycamore tree nearby, and knew there must be a carcass somewhere in the vicinity."[20]

Record Number: 31

Location: Near Fort Ross, CA

Observer(s): William Benitz

Date: 1846–1847

Observation Type: Observation

Reference(s): E. Finley 1937:406

Quotation(s): "Shortly after William Benitz acquired the Fort Ross property he took his rifle and set up the mountain-side to try to kill one of several 'vultures or California condors' perched on the dead limb of a pine tree, in order to obtain feathers, which he knew would be highly prized by his Indian retainers."[21]

Record Number: 32

Location: Mountains of Marin County, CA

Observer(s): Andrew Jackson Grayson

---

20 Lienhard's distinction between vultures and turkey buzzards, along with his location and the time period, makes this a credible sighting.

21 It is unclear whether Benitz ever killed a condor—the Fort Ross Interpretive Association (2001) reports that he did, but documentation is lacking.

Date: July 1847

Observation Type: Observation

Reference(s): Bryant 1891:52–53

Quotation(s): "In the early days of California history it [the California Vulture] was more frequently met with than now, being of a cautious and shy disposition the rapid settlement of the country has partially driven it off to more secluded localities. I remember the time when this vulture was much disliked by the hunter for its ravages upon any large game he may have killed and left exposed for only a short length of time. So powerful is its sight that it will discover a dead deer from an incredible distance while soaring in the air. A case of this kind happened with myself whilst living in the mountains of Marin County, California, in the year 1847. At that time my main dependence for meat wherewith to feed my little family was my rifle. . . . One fine morning I had shot a large and exceedingly fat buck of four points, on the hills above my little cabin. Taking a survey of the sky in every direction I could not discover a single vulture, and, as my cabin was but a short distance from the spot, I concluded not to cover my game as I could return with my horse to pack it home before the vultures would be likely to trouble it. But for this lack of caution I was doomed, as in many other events in my life, to disappointment. I was gone about two hours, when, on returning, I found my game surrounded and covered by a flock of at least a dozen vultures, and others still coming. Some so far up in the heavens as to appear like a small black speck upon the clear blue sky."

Record Number: 33

Location: Mouth of Feather River, CA

Observer(s): Mark Hopkins

Date: 19 September 1849

Observation Type: Killed

Reference(s): Stillman 1967:27

Quotation(s): "Just before night, Mark shot a large bird in the top of a tree, which we thought was a wild turkey. It was directly over our heads, and fell into the water alongside the boat. It measured nine feet from tip to tip of wings, and its head and neck were bare of feathers and of yellow color. It was of the vulture family."

Record Number: 34

Location: Yuba River Canyon, CA

Observer(s): Elisha Douglass Perkins

Date: 20 September 1849

Observation Type: Observation

Reference(s): T. Clark 1967:135

Quotation(s): [Descending through the Yuba River Canyon toward the Sacramento Valley] "Our road was improving rapidly, hills becoming less lofty & steep, country more rolling. Pines & their kind are disappearing & in their place we have scrub oaks. Saw in my mornings march several large vultures, such as I never before met with, perched on a dead tree. Their heads & necks entirely bare of feathers & red—body black, & immense claws & beak. A buzzard by the side of one looked like a black bird, & as one sailed close over my head I judged his spread of wing to be seven foot. I should take them to be a species of condor from the recollections I have of the bird."

Record Number: 35

Location: Sierra foothills, Plumas County, CA

Observer(s): J. G. Bruff party

Date: 20 October 1849

Observation Type: Killed

Reference(s): Reed and Gaines 1949:204

Quotation(s): "6 dead and 1 abandoned ox on road . . . Saw on the road side a small black & yellow fox, dead, also a dead deer, and numerous remains of them. Shot a large very dark brown Vulture, measuring 9 feet from tip to tip."

Record Number: 36

Location: Mill Creek area, Tehama County, CA

Observer(s): J. G. Bruff party

Date: 15–19 November 1849

Observation Type: Observation

Reference(s): Reed and Gaines 1949:240, 245

Quotation(s): *Reed and Gaines 1949, 240 (15 November 1849):* "Poyle and his comrades returned late, unsuccessful. Shot nothing, nor obtained the deer left hanging to a tree; the eagles & vultures had left nothing but the skeleton & skin; an officious and disagreeable interferance with our rights."

Reed and Gaines 1949, 245 (19 November 1849): "Seymours young man assisted, to skin & cut up the meat [of an ox they had killed]. Poyle, bro't a sack full of meat up from the ravine ½ mile down, and we had liver & coffee for dinner—After dinner P. returned for balance of the meat, and reached the spot just in time to save it, as the eagles & vultures were gathering apace, and commenced operations."[22]

Record Number: 37

Location: Yoncalla, Douglas County, OR

Observer(s): Roselle Putnam[23]

Date: 1849–1852

Observation Type: Killed?[24]

Reference(s): Putnam 1928:262

Quotation(s): "The largest wild bird in the country is the vulture which is only an overgrown buzzard—it only preys on the dead carcase—I saw one measured which I think was between ten & eleven feet from the point of one wing to the point of the other."

Record Number: 38

Location: Mill Creek area, Tehama County, CA

Observer(s): J. G. Bruff party

Date: 6–24 February 1850

---

22 The Bruff party had shot a condor earlier in the vicinity (see record 35). This, combined with the fact that Turkey Vultures have generally migrated south by mid-November in northern California, makes these likely condor observations.

23 Roselle Putnam was Jesse Applegate's eldest daughter. Applegate is one of the best known of Oregon's pioneers. He took part in the early government of Oregon and helped establish the Applegate Trail (a safer alternative to boating the Columbia River rapids). The Applegate River (a tributary to the Rogue River) and Applegate Valley of southern Oregon are named in his honor.

24 The bird was measured, but the account does not mention whether it was dead or alive.

Observation Type: Observation

Reference(s): Reed and Gaines 1949:301, 306–8.

Quotation(s): *Reed and Gaines 1949, p. 301 (6 February 1850)*: "3 large vultures passed over."

p. 306 (19 February 1850): "About 4 P.M. I noticed 12 vultures, soaring very high, in circles, over head, and moving toward the S.W."

p. 307 (20 February 1850): "Numerous vultures & eagles flying about."

p. 308 (24 February 1850): "Many eagles & vultures soaring over head.— No doubt attracted by the ox carcasses."[25]

Record Number: 39

Location: Mill Creek area, Tehama County, CA

Observer(s): J. G. Bruff party

Date: 4–29 March 1850

Observation Type: Observation

Reference(s): Reed and Gaines 1949:311, 325.

Quotation(s):

p. 311 (4 March 1850): "Parties returned at Sun-Set, and soon after. They bro't in a deer. Yesterday they shot one, and eat half, hanging the other half in a tree: Today they found but one ham of it, the eagles & vultures had feasted upon it."

p. 325 (29 March 1850): "Saw a bald eagle and several vultures soaring over head."[26]

Record Number: 40

Location: West Branch, North Fork of the Feather River, CA

Observer(s): Samuel Seabough

Date: 1850–1880[27]

25 The Bruff party had shot a condor earlier in the vicinity (see record 35). This, combined with the description of the vultures as "large" and the fact that Turkey Vultures are generally absent in northern California in February, makes these likely condor observations.

26 The Bruff party had shot a condor earlier in the vicinity (see record 35).

27 Seabough arrived in California in 1850 (Cummins 1893), so this individual must have been collected between 1850 and the publication of this article in 1880. No further details are available to narrow the timing of this incident.

Observation Type: Killed

Reference(s): Seabough (1880)

Quotation(s): "The forests of the Sierra Nevada are not remarkable for a great variety of animal life. There is the black-tailed deer, the large gray wolf . . . now and then a vulture almost as large as the Andean condor—I have seen a specimen shot on the West Branch, North Fork of Feather river, that measured 11 feet 6 inches from tip to tip of its outstretched wings."

Record Number: 41

Location: Coast Range, Mendocino County, CA

Observer(s): A. K. Benton

Date: 1854–1856[28]

Observation Type: Killed

Reference(s): *Daily Union* (Sacramento) 1 April 1856

Quotation(s): "Mr. Benton has exhibited to us several of the warlike implements of the various tribes—among them an oval-shaped, sharp-edged flint, which is used as an eating knife in peace, and scalpel in war; also the brace bone of an elk's leg, smoothly rounded , and as sharp as a needle at the point. This weapon our informant seized from a refractory Indian. . . . Last, but not least, of this cabinet of curiosities, is a plume of raven blackness, plucked from a vulture killed in the coast range of mountains. This feather is over two feet long, and the quill barrel measured an inch in circumference. The bird from which it was taken yielded itself reluctantly to his captor, although pierced to the heart by the unerring bullet of the sportsman."

Record Number: 42

Location: Near Chico, CA

Observer(s): Unknown

Date: 1854

Observation Type: Killed

---

28 The dates given for this record are when A. K. Benton was at the Nomen Lackee (Round Valley) Indian Reservation. From the quotation in the newspaper it sounds like Benton was the sportsman who shot the condor, but it is not entirely clear.

Reference(s): *Daily Union* (Sacramento) 21 June 1854

Quotation(s): "A CALIFORNIA VULTURE.—The Editor of the Marysville *Herald* may well 'plume himself', on receipt of a vulture's quill measuring twenty-five inches in length. The bird measured nine feet four inches from tip to tip of the wings. It was shot near Chico."

Record Number: 43

Location: Near Fort Vancouver, WA

Observer(s): J. G. Cooper

Date: January 1854

Observation Type: Observation

Reference(s): Cooper and Suckley 1859:141; Belding 1890:25

Quotation(s): *Cooper and Suckley 1859, p. 141:* "In January, 1854, I saw, during a very cold period, a bird which I took for this [the California Vulture], from its great size, peculiar flight, and long bare neck, which it stretched out as it sat on a high dead tree, so as to be scarcely mistakable for any other bird."

Belding 1890, p. 25: "Cooper, 1870. This confirms the observations of Dr. Suckley and myself, as we saw none [California Vultures] during a long residence and travels near the Columbia, except one which I supposed to be this, seen at Fort Vancouver in January. Like several other birds seen there by Townsend and Nuttall, they seem to have retired more to the south since 1834."

Record Number: 44

Location: American River, El Dorado County (near the store of Woods & Kenyon), CA

Observer(s): Unknown

Date: March 1854

Observation Type: Killed

Reference(s): *Daily Union* (Sacramento) 11 March 1854

Quotation(s): "CALIFORNIA VULTURE.—A vulture of enormous proportions was shot on the American river, near the store of Woods & Kenyon, in El Dorado county, a few days since, which measured nine feet from tip to tip of its wings. A friend presented us yesterday with a quill, which is a quill from one of its wings, with the remark that it was handed us as a weapon with which to defend the rights of the people. We shall endeavor to apply it to that purpose."

Record Number: 45

Location: South Fork of American River (North Canyon), CA

Observer(s): Alonzo Winship and Jesse Millikan

Date: Fall 1854

Observation Type: Live-captured and released

Reference(s): Millikan 1900:12–13

Quotation(s): "As he [Jesse Millikan] was crossing the aqueduct over North Canyon, he saw an enormous condor asleep at the base of a cliff that jutted about twenty feet above the flume. Surprised that the bird had not been awakened by his footsteps along the flume, he hesitated a moment, then decided to attempt to kill the bird. Having nothing but his shovel he threw it with all his force, striking the condor and breaking its wing."[29]

Record Number: 46

Location: Vicinity of the redwoods of Contra Costa, CA

Observer(s): Joseph P. Lamson

Date: January–February 1854

Observation Type: Killed (x4); 1 wounded and escaped; other observations

Reference(s): Lamson 1852–1861:283; Lamson 1878:152–54

Quotation(s): *Lamson 1852–1861, 283*: "Thursday 2 February [1854]. I was standing at the door of my cabin, which I heard the report of a rifle, and turning my eyes in the direction of the sound, I saw a California Vulture fall to the ground. I hastened up the cañon, and speedily purchased the bird of the owner, who did not place a very high value on it. It was not a large one, being only eight feet six inches in alar extent, and three feet ten inches long from the point of the bill to the end of the tail,

---

29 The story goes on to describe in great detail how the wounded condor was live-captured and given away to the owners of a slaughterhouse, who kept it in a corral, and how it then disappeared, was recaptured, and ultimately escaped.

and weighing twenty-one pounds. The longest feathers in the wings were twenty-six inches. I was desirous of preserving the skin of one of these birds, but chose to wait until I should obtain a larger specimen. So I cut off the wings, and pulled out some of the feathers from the body, which I preserved as well as some of the bones and threw the body away."

Lamson 1878, 152–54: "February 9, 1854. In a walk some days since through the Redwoods, I encountered an old man by the side of the road engaged in making shingles. He was a very coarse-looking fellow with a dark complexion and a black, bushy beard, that more than half covered his face, giving an additional grimness to his rough, harsh features. He was an old Kentucky rifleman, and, as I learned to-day, a first-rate marksman. He had shot a Vulture some time before, and it was lying near his cabin, half decayed. Some quills were scattered over the ground, and I picked up two or three of them, when he ordered me in the rudest manner to leave them. I then offered to buy some of them, but he would neither sell nor give them away. He wanted them for himself. . . . To-day I passed his cabin again, and he accosted me with considerable civility . . . he had shot two Vultures yesterday, though one of them, which he had only wing-tipped, and tied to a stake, had escaped. He was willing to sell me the remaining bird, and the payment of five bits made me its owner. . . . I skinned my bird, and left it with the Kentuckian, while I continued my walk. . . . I saw six or eight of them perched on trees, sitting in perfect idleness and scarcely moving. . . . On returning, I called for the skin of my bird which measured nine feet four inches from tip to tip of the wings, and three feet eleven inches in length."[30]

Record Number: 47

Location: Vicinity of the redwoods of Contra Costa, CA

Observer(s): Alexander S. Taylor

Date: 1855

Observation Type: Killed

Reference(s): California Academy of Natural Sciences 1863:71

Quotation(s): *"Donations to the Cabinet.* From Mr. A. C. Taylor, quills taken from a California Vulture (*Cathartes californianus,* Shaw) killed in the vicinity of the Red Woods of Contra Costa. The bird measured 13½ feet across the wings[.]"

---

30 Additional details of this story are provided in Lamson's journal (Lamson 1852–1861).

Record Number: 48

Location: Sacramento Valley, CA

Observer(s): John Strong Newberry

Date: July 1855

Observation Type: Observation

Reference(s): Newberry 1857:73

Quotation(s): "A portion of every day's experience in our march through the Sacramento valley was a pleasure in watching the graceful evolutions of this splendid bird [the Californian Vulture]. Its colors are pleasing; the head orange, body black, with wings brown and white and black, while its flight is easy and effortless, almost beyond that of any other bird. As I sometimes recall the characteristic scenery of California, those interminable stretches of waving grain, with, here and there, between the rounded hills, orchard-like clumps of oak, a scene so solitary and yet so home-like, over these oat-covered plains and slopes, golden yellow in the sunshine, always floats the shadow of the vulture."

Record Number: 49

Location: Siskiyou Mountains, Klamath Basin, CA

Observer(s): John Strong Newberry

Date: August 1855

Observation Type: Observation

Reference(s): Newberry 1857:73

Quotation(s): "After we left the Sacramento valley, we saw very few [Californian Vultures] in the Klamath basin, and none within the limits of Oregon. It is sometimes found there, but much more rarely than in California."

Record Number: 50

Location: Mendocino County, CA

Observer(s): Lyman Belding

Date: 1856–1878

Observation Type: Observation

Reference(s): Belding 1878, in litt.; Fisher 1920[31]

Quotation(s): "I have never shot a Cal Condor have seen a few along Feather River [probably in the vicinity of Marysville—see record 51] in former years & a few in Mendocino Co - - I will try to get you one or two - have not much confidence in my ability to do so."

Record Number: 51

Location: Near Marysville, Yuba County, CA

Observer(s): Lyman Belding

Date: 1856–1879

Observation Type: Observation

Reference(s): Belding 1879:437; Fisher 1920

Quotation(s): "The California Condor appears to be very rare in this region. I have seen it on no more than two or three occasions in Yuba County in winter, and do not think I have seen it at any other place. They probably visit the vicinity of Marysville only in winter, and are never common."

Record Number: 52

Location: Near Sacramento (caught on Mrs. Harrold's ranch), CA

Observer(s): Unknown

Date: September 1857

Observation Type: Live-captured

Reference(s): *Daily Union* (Sacramento) 24 September 1857

Quotation(s): "DECIDEDLY VORACIOUS.— Mr. Sutton, of the Western Hotel, corner of K and 10th streets, was presented a few days since with a young vulture, which he has placed in the yard of his establishment. . . . Length of wings from tip to tip, about 10 feet 6 inches; length of head and beak, 7 inches; length of claws, from 7 to 9 inches. He is fed regularly and literally on raw heads and bloody bones, and can clean a skull or bone in the most approved style. . . . The vulture was caught on Mrs. Harrold's ranch, near this city [Sacramento]."

---

31 Fisher (1920) provides information on when Belding was in California, which helps narrow down the timing of this observation.

Record Number: 53

Location: Lower Napa Valley, CA

Observer(s): Frank A. Leach

Date: 1857–1860

Observation Type: Observation

Reference(s): Leach 1929:23

Quotation(s): "In the later [18]50's, in the central and northern parts of the state, it was not uncommon also to see the great Condors (*Gymnogyps californianus*) associated with flocks of a dozen or more buzzards, feeding on the remains of a dead horse or steer. I frequently saw them between the years of 1857 and 1860 on the bare hills of lower Napa Valley. They were so much larger than the buzzards that there was no trouble in distinguishing one from the other. Generally where there was a flock of the smaller birds gathered about a carcass, there would be two or three of the big Condors. It is my impression that after 1859 or 1860 the latter were seldom seen, in the Napa section at least; and I think the extinction of the Condor in northern California took place in the decade following 1860."

Record Number: 54

Location: Pope Valley, near Saint Helena, Napa Valley, CA

Observer(s): J. B. Wright

Date: January 1858

Observation Type: Killed

Reference(s): *Daily Alta California* (San Francisco) 4 February 1858

Quotation(s): "Vulture Shot.—Last week, while Mr. J. B. Wright, of Pope Valley, near St. Helena, Napa Valley, was out hunting, he shot a large vulture, that was flying off with a hare it had killed, weighing nine pounds. The bird measured fourteen feet from tip to tip of wings. We have one of the tail feathers in our office, that measures twenty-six inches in length."

Record Number: 55

Location: Mouth of Fraser River, BC

Observer(s): John Keast Lord

Date: 1858–1866

Observation Type: Observation

Reference(s): Lord 1866:291

Quotation(s): "Catha[r]tis californianus . . . Mouth of Fraiser River. Seldom visits the interior."

Record Number: 56

Location: Russian River area of the Coast Range, CA

Observer(s): L. L. Davis

Date: 1861

Observation Type: Killed

Reference(s): *Daily Union* (Sacramento) 18 June 1861

Quotation(s): "A LARGE BIRD.—The Grass Valley *National* has the following account of a California vulture: We write this article with a pen plucked from the pinions of a vulture killed on the coast range, which weighed thirty pounds, and measured from tip to tip of its wings, fourteen feet. The longest quill measured thirty-four inches in length. L. L. Davis who presented us with the quill informs us that the vulture is quite common on the Russian river portion of the Coast Range. They are very large, and particularly fond of pork. They will descend with a sweep upon a forty-pounder, kill him at the first blow, seize him in their talons and bear him away with scarcely any perceptible hindrance to their flight. Davis assures us that they would carry off children if any were to be found."

Record Number: 57

Location: Plumas County, CA

Observer(s): S. Stevens

Date: 1865

Observation Type: Killed

Reference(s): *Daily Union* (Sacramento) 25 November 1865

Quotation(s): "A HUGE BIRD.—We saw recently, says the Nevada *Transcript*, at the shop of Z. Davis, the wing of a bird recently shot by S. Stevens in Plumas county. The feathered giant is said to be of the Condor family, and measured eleven feet from tip to tip of the wings."

Record Number: 58

Location: Near Marin County paper mills, CA

Observer(s): Unknown

Date: 17 August 1868

Observation Type: Killed

Reference(s): *Bulletin* (San Francisco) 19 August 1868

Quotation(s): "PROUD BIRD OF THE MOUNTAIN.—A condor was shot on Monday near the Marin county Paper Mills, which measured nine feet from tip to tip of wings. It was brought to the city yesterday, and will be taken to a taxidermist to be stuffed and preserved as a curiosity."

Record Number: 59

Location: Winnemucca, NV

Observer(s): Unknown

Date: August 1871

Observation Type: Observation

Reference(s): *Daily Union* (Sacramento) 26 August 1871

Quotation(s): "A MONSTER.—Last Tuesday evening, about 7 o'clock, says the Winnemucca 'Register' of August 19th, the people in the lower town were startled by the sudden appearance of a huge monster we are at a loss to know whether to call fowl or beast, notwithstanding it had wings and could fly. It was certainly the biggest creature ever seen in this country with feathers. If a bird, it belongs to a giant species unknown to American ornithology. Our attention was attracted by hearing some one sing out 'holy mother, see that cow with wings.' We stepped to the door just in time to see the monster alight with something of a crash on the roof of Mrs. Collier's dwelling-house, where it remained for several minutes taking a quiet survey of the land and the astonished multitude who stood gazing at that unexpected visitor. It could not have weighed less than 75 or 100 pounds, with a pair of ponderous wings, which, when stretched out to the breeze, must have been fully 12 feet from tip to tip. Its color was that of a raven, with the exception that the tips of its wings and tail were white. An 'old salt' who happened to get sight of the bird thinks he must be a renegade member of the condor family. He says he has frequently met with just such 'critters' on the coast of South America."

Record Number: 60

Location: Mendocino County, CA

Observer(s): Unknown

Date: Winter 1872

Observation Type: Killed

Reference(s): *Tribune* (Chicago) 24 February 1872

Quotation(s): "A vulture eagle, measuring upward of nine feet from tip to tip, has been killed in Mendocino County, Cal."

Record Number: 61

Location: Salmon Creek, Marin County, CA

Observer(s): Julius Poirsons

Date: February 1873

Observation Type: Killed

Reference(s): *Daily Evening Bulletin* (San Francisco) 19 February 1873

Quotation(s): "A California condor was killed in Marin county the other day which measured nine feet from tip to tip and weighed seventeen pounds."[32]

Record Number: 62

Location: Near the peak of Mount Shasta, CA

Observer(s): Benjamin P. Avery

Date: September 1873

Observation Type: Observation

Reference(s): Avery 1874:476

Quotation(s): "A few snowbirds were twittering a thousand or two thousand feet below, and nearly up to the very crest of the Main Peak [of Mount Shasta] we saw a solitary California vulture wheeling slowly around."

---

32 Several other accounts of this shooting were given in several other newspaper articles and publications around the same time: *Daily Union* (Sacramento) 19 February 1873; *Daily Evening Bulletin* (San Francisco) 4 March 1873; Proceedings of the California Academy of Sciences 5:43; *Bulletin* (San Francisco) 22 March 1873.

Record Number: 63

Location: Near the hot springs above Boise City, ID

Observer(s): General T. E. Wilcox

Date: Fall 1879

Observation Type: Observation

Reference(s): Lyon 1918:25

Quotation(s): "In the fall of 1879 I came upon two [California Vultures] which were feeding on the carcass of a sheep. They hissed at me and ran along the ground for some distance before they were able to rise in flight. They were much larger than turkey buzzards, with which I was quite familiar, and I was very close to them so that I could not be mistaken in their identity. The cattle-men said that the California vulture or buzzard was not uncommon there before they began to poison carcasses to kill wolves. . . . Boise River mountains rise to over 7000 feet just back of where the vultures were feeding. The exact locality was near the Hot Springs above Boise City. Poison and population have now destroyed that far northern habitat."

Record Number: 64

Location: Foothills southwest of Mount Lassen, CA

Observer(s): Unknown

Date: 1879–1884

Observation Type: Observation

Reference(s): C. Townsend 1887:201

Quotation(s): "In 1884 a hunter at Red Bluff told me that he had killed a vulture of immense size in the southeastern part of Tehama County two or three years previous, and that he had seen others in the foothills southwest of Mount Lassen within the last four or five years. As this is all the information I could obtain with regard to this species, it has probably almost disappeared from Northern California, where it was once certainly common."

Record Number: 65

Location: South Eel River, Humboldt County, CA

Observer(s): Mr. Adams

Date: Spring 1880

Observation Type: Killed

Reference(s): *Bulletin* (San Francisco) 5 April 1880

Quotation(s): "On South Eel river, Humbold[t] county, Mr. Adams recently poisoned a bird of the vulture species which measured nine feet across the wings, four feet from beak to tail and eight inches from crown to tip of beak."

Record Number: 66

Location: Reeds Creek Canyon, Tehama County, CA

Observer(s): John Bogard

Date: May 1880

Observation Type: Killed

Reference(s): Daily Evening Bulletin (San Francisco) 7 May 1880

Quotation(s): "A large California vulture was killed a few days ago in Reed's creek canyon, Tehama county, by John Bogard of Tehama. It measured 8 feet from tip to tip."

Record Number: 67

Location: Burrard Inlet, BC

Observer(s): John Fannin

Date: September 1880

Observation Type: Observation

Reference(s): Fannin 1891:22

Quotation(s): "In September, 1880, I saw two of these birds [California Vultures] at Burrard Inlet. It is more probable they are accidental visitants here."

Record Number: 68

Location: Mountains south of Mount Lassen, CA

Observer(s): Unknown

Date: 1881–1882

Observation Type: Killed

Reference(s): C. Townsend 1887:201

Quotation(s): "In 1884 a hunter at Red Bluff told me that he had killed a vulture of immense size in the southeastern part of Tehama County two or three years previous, and that he had seen others in the foothills southwest of Mount Lassen within the last four or five years. As this is all the information I could obtain with regard to this species, it has probably almost disappeared from Northern California, where it was once certainly common."

Record Number: 69

Location: Vicinity of San Rafael, CA

Observer(s): Unknown

Date: 1882 or before

Observation Type: Live-captured

Reference(s): Gassaway 1882:89[33]

Quotation(s): "Attached to one of the many picnic 'groves' is a local celebrity of unique accomplishments. He is called the 'San Rafael Octopus,' and is so designated on account of the facility with which he hugs eight girls at a time, and renders himself generally useful to visiting organizations. A young man gifted like that ought to make a proud record for himself at an Oakland church sociable. Why the picnic ground of the period should not be considered complete without an attenuated, disreputable and anything but inodorous bear chained to a post in its midst, as well as a melancholy eagle moping in a chicken coop, it would be hard to tell. In addition to these forlorn captives, we saw at one place a huge vulture, or California condor, tied to a stake. The proprietor kindly offered to illustrate this bird's proverbial voracity by feeding it with fish. After eagerly devouring its weight in tomcods three times over, it paused

---

33 This is a compilation of columns originally published in the San Francisco Evening Post, written circa 1880–1882.

to gasp for breath with the tail of the last fish sticking out of its stuffed and swollen neck."

Record Number: 70

Location: Chico, Butte County, CA

Observer(s): William Proud

Date: Prior to 1890, probably 1880s[34]

Observation Type: Observation

Reference(s): Belding 1890:24

Quotation(s): Belding quotes Mr. William Proud as noting that condors are "sometimes seen" near Chico, California.

Record Number: 71

Location: Lulu Island (Fraser River delta), BC

Observer(s): Mr. W. London

Date: 1888–1889

Observation Type: Observation

Reference(s): Rhoads 1893:39

Quotation(s): "Seen on Lulu Island as late as 'three or four years ago' by Mr. W. London.[35] 'None seen since, used to be common.'"

Record Number: 72

Location: Kneeland Prairie, Humboldt County, CA

Observer(s): Unknown

Date: 1889 or 1890

Observation Type: Killed; specimen in the Clarke Museum in Eureka, CA

Reference(s): F. Smith 1916:205

---

34 All of William Proud's observations of other birds reported in Belding (1890) were from 1884–1885.

35 William London was one of the first white settlers on Lulu Island, arriving in 1881. He was a prominent businessman and served as a city councillor for the City of Richmond (Lulu Island) from 1883 to 1887 (Kidd 1927).

Quotation(s): "There is no doubt but that the Condor (*Gymnogyps californianus*) once occurred in numbers in Humboldt County, California. There are now two mounted specimens in Eureka. One, in the collection of the Public Library, was mounted by Mr. Charles Fiebig, and was secured from a dead spruce tree on the Devoy place, on Kneeland prairie, eighteen miles from Eureka, altitude 2200 feet, in the fall of 1889 or 1890."

Record Number: 73

Location: Yager Creek, Humboldt County, CA

Observer(s): F. H. Ottmer

Date: Fall 1892

Observation Type: Killed; specimen in Eureka High School's Hall of Ornithology, Eureka, CA

Reference(s): F. Smith 1916:205

Quotation(s): "The [California Condor specimen] is in the collection of Dr. Ottemer in Eureka and was mounted by William Rotermund. This specimen was captured near the old Olmstead place on Yager Creek, altitude 1800 feet, about sixty miles east of Eureka, in the fall of 1892. Old settlers claim that the Condor was plentiful in early days in the Humboldt region. In my opinion it is now [in 1916] extinct here."

Record Number: 74

Location: A few miles east of Coulee City, WA

Observer(s): C. Hart Merriam

Date: September 1897

Observation Type: Observation

Reference(s): Jewett et al. 1953:160

Quotation(s): "The last record of the species [California Condor] for the state [Washington] appears to be that of Dr. C. Hart Merriam (letter of January 4, 1921). In the early morning of September 30, 1897, Dr. Merriam saw a condor on the ground in open country a few miles east of Coulee City."

Record Number: 75

Location: Southern coast, OR

Observer(s): Henry Peck

Date: Probably late 1800s

Observation Type: Killed

Reference(s): W. Finley 1908b:10

Quotation(s): "Mr. [Henry] Peck also gives the record of a condor that was killed on the coast of southern Oregon a number of years ago."

Record Number: 76

Location: Curry County, OR

Observer(s): Unknown

Date: Prior to 1900

Observation Type: Observation

Reference(s): Koford (1941) notes, 11 April (at the Museum of Vertebrate Zoology, Berkeley, CA)

Quotation(s): "One rancher had told him [Stanley Jewett] that after they started poisoning for varmints the vultures disappeared but the condors did not. Jewett asked him what he meant by 'vulture' and the man gave an excellent condor description."[36]

Record Number: 77

Location: Mountains north of San Francisco, Marin County, CA

Observer(s): Unknown

Date: 1900–1905

Observation Type: Killed

Reference(s): Chicago Field Museum specimen 39613

Quotation(s): None

---

36 The rancher apparently got Turkey Vultures and condors reversed at first, which Jewett clarified by asking him what he meant by "vulture."

Record Number: 78

Location: Near Drain, OR

Observer(s): George D. Peck and Henry Peck

Date: 1 June 1903

Observation Type: Observation

Reference(s): Peck 1904:55; W. Finley 1908b:10; Gabrielson and Jewett 1970:181

Quotation(s): *Peck 1904, 55*: "June 1, 1903, I saw two Cal. Vultures. They were at a great height and I could not have identified them if I had not often seen them in Los Angeles County, Cal. I saw several of the great Vultures during the month of June. The birds that I saw were about thirty miles from the coast."

W. Finley 1908b, 10: "Mr. Henry Peck informs me that on or about July 4, 1903, he and his father saw two California condors at Drain, Douglass County, Oregon. They were quite high in the air and were sailing about over the mountains. The elder Mr. Peck saw them several times after that. He states the birds were instantly recognized by both of them."

Gabrielson and Jewett 1970, 181: "Jewett has talked to several well-informed woodsmen who described accurately to him condors seen in southern Oregon at about the time of the Peck observation, and it seems highly probable that two or more of these big birds strayed into southern Oregon, perhaps to remain for some time."

Record Number: 79

Location: Near Drain, OR

Observer(s): Henry Peck

Date: 9 March 1904

Observation Type: Observation

Reference(s): W. Finley 1908b:10; Gabrielson and Jewett 1970:181

Quotation(s): *W. Finley 1908b, 10*: "In March 1904, Mr. Henry Peck writes, 'I saw four condors which were very close to me, almost within gun shot. I recognized them first by their size, and second by the white feathers under their wings. The birds were all flying very low, as there was a high wind blowing.'"

Gabrielson and Jewett 1970, 181: "[Stanley] Jewett has talked to several well-informed woodsmen who described accurately to him condors seen in southern Oregon at about the time of the Peck observation, and it seems highly probable that two or more of these big birds strayed into southern Oregon, perhaps to remain for some time."

Record Number: 80

Location: Kibesillah (near Fort Bragg, Mendocino County), CA

Observer(s): Miss Cecile Clarke

Date: Fall 1912

Observation Type: Observation

Reference(s): Clarke 1971, in litt.

Quotation(s): In a March 1971 letter to Sanford Wilbur, Miss C. Clarke wrote: "In the fall of 1912, I saw one [California Condor] flying south in Kibesillah. I did not know the bird, but I was up on eagles, hawks, turkey vultures, etc. [and so] I decided to find out. I asked everyone I knew and finally a very old man told me that he had seen it. He said it was the same old condor that goes north in the spring and south in the fall."

Record Number: 81

Location: Siskiyou County, CA

Observer(s): Henry Frazier

Date: Spring 1925

Observation Type: Observation

Reference(s): Henry H. Frazier, Morro Bay, CA, pers. comm. with Sanford R. Wilbur, 20 February 1971

Quotation(s): "In the spring of 1925, on the divide west of Copco Dam in Siskiyou County, Frazier saw six birds that he thought were adult California condors. There was snow on the ground, and the six birds were feeding on a dead cow. Mr. Frazier gave a good description of condors."

Index

Numbers in bold type refer to illustrations.